Contributions to Management Science

For further volumes:
http://www.springer.com/series/1505

Andreas Kemper

Valuation of Network Effects in Software Markets

A Complex Networks Approach

Physica-Verlag

Dr. Andreas Kemper
Azaleenweg 114
60433 Frankfurt
Germany
andreas.kemper@googlemail.com

ISBN 978-3-7908-2813-9 e-ISBN 978-3-7908-2367-7
DOI: 10.1007/978-3-7908-2367-7
Springer Heidelberg Dordrecht London New York

Diss., European Business School (EBS), 2009, 154

Cover design: SPi Publisher Services

Printed on acid-free paper

Physica-Verlag is a brand of Springer
Springer is part of Springer Science+Business Media (www.springer.com)

Preface

This book is the product of my research at the Endowed Chair of Corporate Finance and Capital Markets at the European Business School. Many people devoted their inspiration, time, and knowledge in order to support me with my interdisciplinary research. Thank you all for your support.

Especially, I thank my academic supervisor Prof. Ulrich Hommel, Ph.D., who provided me with the freedom to pursue my research interests and who always supported me generously. In addition, I am very grateful to my co-supervisor Prof. Dr. Susanne Strahringer (TU Dresden), who graciously accepted the task of refereeing my book. Also, I thank Prof. Dr. Matthias Krause of the Chair for Theoretical Computer Science in Mannheim for the technical and computer scientific support.

During my research visits I received generous support from Prof. Yaneer Bar-Yam (NECSI and Harvard) and Prof. John D. Sterman (MIT), who introduced me to the field of System Dynamics and Complex Systems, as well as from Prof. Victor Mossotti (U.S. Geological Survey), Prof. Hiroki Sayama (NECSI and Binghamton), Prof. Markus de Aguiar (Universidade Estadual de Campinas), Prof. Michel Baranger (MIT), Prof. Jay W. Forrester (MIT), and Prof. Brian D. Josephson (Cambridge). Thank you very much for numerous inspiring discussions.

Many other people supported the development of my ideas. Accordingly, I am particularly grateful that I had Dirk Stegemann, Alexander Becker, Dr. Alexander Falkenberg, Oliver Gaberle, Frederik Kruska, Dr. Benjamin Gansel, Dr. Matthias Eder, Ronja Hoelzer, Dr. Michaela Stollfuss, and Dr. Martin Hülsmann at my side with whom I could not only share scientific ideas but also various social events. These extraordinary friends have been a source of inspiration for me in many ways, often far beyond scientific questions.

Furthermore, I thank the CORPFIN team (ebs), Gudrun Fehler (ebs), Dr. Sigrid Berka (MIT), Mae NGuyen (CIA), Dr. Ben Goertzel (Biomind), Dr. Richard Metzler (NECSI and MIT), Daniel Rosenbloom (NECSI and Harvard), Cherry Ogata, and the whole NECSI team. In addition, I thank Dr. Erik Rauch (NECSI and Princeton), who has left us far too early.

Last but not least, I would like to thank my wife Janna, my son Darius Raphael, my parents Wolfgang and Monika, Melanie, Alexander, Stefanie, and Lara Marie, and the rest of my family who have always supported me and to whom I devote this book.

Contents

List of Figures

List of Tables

Part I
Introduction

Chapter 1
Background and Motivation

> "What's a customer worth? The company that can answer this question precisely is the company with an edge in the customer-based, technology- and information-intensive economy of today. But how can an asset as intangible as customer value be measured?"
>
> (Blattberg et al. 2001)

In this world of constant changes, software markets are no exception. Growing transaction volumes of the merger-driven industry and capital markets pressure to accurately value companies increases the demand for reliable valuation in software markets. This trend is supported by an increasing number of internal capital budgeting decisions that required to determine the financial impact of strategic investments on tangible and intangible assets (Srivastava et al. 1999). Valuation in software markets has always been challenging. The reason for this challenge is their *dynamic nature* which is characterized by exponential growth and decay, fierce competition and highly lucrative rents. Such factors increase the danger of false evaluation and misinterpretation.[1] During the Internet hype the volatility of the stocks reached a peak as software companies were among the best performing stocks, but also among the losers of the subsequent downturn.[2] This volatile development caused great concerns, but financial research was not capable of explaining it despite of considerable research progress as a variety of problems have not yet been resolved (Busse von Colbe 1957; Ballwieser 1987; Moxter 1991; Copeland et al. 1996; Ross et al. 1996; Brealey and Myers 1996; Achleitner and Nathusius 2004). Conventional approaches tend to ignore the following phenomena which are particularly relevant to valuation in software markets: (Rohlfs 1974; Wiese 1990; von Westarp 2003).

1. *Interacting Demand Curve*. Conventional valuation models are frequently based exclusively on the number of actual customers. But consumption decisions in network effect markets depend also on the behavior of potential customers in customer networks.

---

[1] A popular example is that of Microsoft dominating the software market for operating systems, with a reported 2.790 percent increase in profits and a 1.665 percent increase in revenues between 1990 and 1999.

[2] The German new market peaked at a maximum of 9.632 points on March 10th, 2000 and collapsed to 313 points in October, 2002.

A. Kemper, *Valuation of Network Effects in Software Markets*,
Contributions to Management Science, DOI 10.1007/978-3-7908-2367-7_1,

2. *Endogenous Growth*. Growth is assumed to be exogenous in conventional approaches. Hence, endogenous growth is not explained and frequently not even considered.
3. *Divisibility of Goods*. As it is not possible to purchase a fraction of a software package, the assumption concerning the divisibility of goods is violated.
4. *Monopolistic Market Concentration*. Software markets are governed by economies of scale and reinforcing network effects, which cause constant price pressure, dropping results and monopolization.

The most significant deficiency in conventional valuation approaches, however, is that the values of intangible assets are hardly reflected. Traditional approaches fail to account for the value of managerial flexibilities in customer networks, although these can be very valuable in dynamic software markets (Trigeorgis 1996, p. 243) (Hommel and Pritsch 1999, p. 127). Consequently, management tends to focus on cost reduction strategies and investors tend to limit their investments. This behavior, in turn, may lead to missed investment opportunities as financial decisions are based on strategic reasoning rather than on the results of investment analyzes (Hommel and Pritsch 1999). It is the purpose of this research to *explore new venues* to these unresolved research problems.

The subsequent research is driven by the hypothesis that the value of customer networks has to be taken into consideration in order to deliver more stable and relevant valuation results than conventional metrics (Gupta et al. 2001). In this context, research on *network economics* provides additional insights. It stresses that customer relationships are a source of flexibility that have both direct and indirect impact on demand in software markets (Blattberg and Thomas 1998). Such indirect phenomena resulting from the customer network are termed *network effects* (Shapiro and Varian 1998; Katz and Shapiro 1985). Such network effects are endogenous explanations for growth and are of particular interest if a company needs additional investments to cross a critical threshold of users. In such a case, investors have to decide whether the company is liquidated or whether additional financial resources are provided based on the expected benefit from an asymmetric increase of the company value once the critical mass of customers is reached (Richins 1983; Farrell and Saloner 1986; Reichheld 1999). Therefore, it is necessary to outweigh the growth potential with the respective costs as an ignorance of such intangible values is likely to result in inefficient investment decisions. Empirical investigations of listed innovative companies reveal that such intangibles can account for more than 70 percent of the company market value (Srivastava et al. 1998). This underlines the importance of customer networks and fosters a *customer network-centric perspective* on valuation in software markets.

Thus, it is reasonable to disaggregates cash flows on the level of individual customers, such as purchase, cross selling and word-of-mouth activities. Consequently, the combined values of all current and potential customers yield the value of the customer network. From a financial perspective, the managerial flexibilities resulting from such customer networks can be interpreted as *real options*. They capture the upside potential of future cash flows in case of a successful product diffusion process, which depends on the probability of reaching a critical mass of customers within a certain time horizon with the available financial resources. Hence, the following research efforts aim at the core problem of this valuation approach: The derivation of the probability to reach a *critical mass of customers* in a specified time horizon with limited financial resources and the respective volatility of implied cash flows. A solution to this problem requires a solid understanding of the characteristics and dynamics of customer networks in software markets in order to quantify the respective probabilities. For this purpose a network perspective is pursued. Analytical and numerical network models are designed that allow one to conduct quantitative investigations of the adoption and diffusion behavior in customer networks of software markets. A comparison of such approaches reveals that properties, topologies and dynamics of customer networks are the key to an advanced understanding of customer networks and dynamics.

In essence, customer networks are main value drivers of software companies and network effects are main drivers of customer networks. Hence, it is necessary to increase the understanding of customer networks in order to enhance the understanding of valuation in software markets. For this reason the increasing body of research on *complex networks* is a promising approach as it revealed a variety of relevant insights on complex networks in biology and theoretical physics, but also in the social sciences such as information networks, scientific collaboration, epidemiology and communication networks (Albert 2001; Barabasi et al. 2002; Boccara and Cheong 1992). Accordingly, customer networks are represented by Bayesian networks and the economic success of software products is described by a network adoption and diffusion model. Based on such network theoretical descriptions of structural and locational properties of customer networks, their network dynamics are investigated in order to understand consumption interdependencies, market penetration, adoption, and standardization processes in software markets. Various *topologies of networks*, i.e. structures of networks, are investigated with respect to their impact on the diffusion processes in networks. The insights are then used in order to develop a *complex networks adoption and diffusion model of software markets* that facilitates increasing the market transparency by identifying, quantifying, and valuing the optional value of customer networks. It is important to note that it is not possible within the scope of this paper to explore all details. Instead, this book seeks to integrate network effects based on a complex networks approach into a customer network-centric framework for valuation in software markets that allows to increase the quality of valuations, and that of subsequent strategic or investment decisions.

1.1 Research Objectives

The research efforts are primarily aimed to provide innovative insights into valuations in software markets from a network theoretical perspective by accounting for network effects, network topologies and network dynamics and to foster thereby the understanding of relations in customer networks as valuable flexibilities that can be valued as real options. The *overall research goal* is to overcome existing limitations of valuation research and thereby to increase the quality of valuations in software markets by adopting a network perspective that allows an identification, quantification and valuation of network effects through a deeper understanding of network characteristics and their dynamics. This primary objective comprises the following subgoals.

1. *Identification of Network Effects in Software Markets.* The first research objective is to identify the role of network effects in software markets. Based on a literature review, the underlying hypothesis is that network effects can be vital determinants of software market dynamics which are currently not adequately represented. It is assumed that filling this research gap increases the transparency of valuations in software markets.
2. *Valuation of Network Effects in Softwares Markets.* After an identification of the network effects, the main objective of this research is to integrate them into software market valuations. Hence, it is necessary to determine the optimal valuation approach for software market valuations that accounts for network effects. Then, these insights are integrated with those on valuation. As research on network effects and complex networks cross-fertilized several other research areas, the respective network data is assumed to enhance the quality of valuations.
3. *Quantification of Network Effects in Softwares Markets.* The key to the integration of network effects of customer networks into valuation in software markets is the quantification of such effects. As software market models are successfully applied to derive additional insights into the structure and dynamics of software markets, it is suggested that they also provide the potential to quantify network effects in software markets.
4. *Valuation Framework for Valuations in Software Markets.* Once the previous research objectives are achieved, it is desirable to consolidate the derived insights. Hence, the next goal is to develop a respective framework for customer network-centric valuation in software markets that accounts for network effects.
5. *Network Topologies and Network Dynamics in Software Markets.* A reconsideration of the developed framework pinpoints to the vital importants of network properties and dynamics. While conventional research frequently assumes homogeneous random customer networks, complex networks research provides an array of concepts and insights on network properties, topologies and dynamics.[3] Hence, a further objective is to refine the developed valuation approach by accounting for the respective complex networks insights and to resolve the most relevant open aspects related to this approach.
6. *Bridging the Research Gap between Social Sciences and Natural Sciences.* Despite of first progress there is a considerable gap between various schools of thought although there are also numerous examples indicating the cross-fertilization of interdisciplinary research. Hence, it is a final research objective

[3] Please confer Sect. 10.3.

to investigate the opportunities and limitations of the interdisciplinary research approach, in order to outline new venues to bridge the existing gap particularly between the social sciences and natural sciences.

After this outline of the research objectives, the design of the research is depicted in the subsequent section.

1.2 Research Design

Research can be classified into three different types of approaches, i.e. the formal-analytic research strategy, the empirical research approach and the content-analytic research strategy (Grochla 1978).

1. *Formal-Analytic Research Strategy*. Formal-analytic research is conducted based on simplifying abstracts of problems that are applied in order to design decision relevant models that allow one to deduce quantifiable solutions to a problem.
2. *Empirical Research Strategy*. Empirical research strategies aim at formulating and testing hypotheses based on cause-and-effect relationships. A prerequisite is a testable hypotheses with respect to the research object.
3. *Content-analytic Research Strategy*. The content-analytic strategy is based on a structured analysis of complex interdependencies as it allows one to identify and extend the plausibility material.

In the following, we present an overview on the approaches that are pursued in this book. First, a content-analytic approach serves as a basis for the theoretical framework in which complex interdependencies between software markets, valuation and network effects are investigated.[4] The investigations are primarily based on plausible deductions and empirical indications. In a next step, a formal-analytic Markov matrix diffusion model is developed in order to analyze the principle mechanics of product diffusion processes in software markets.[5] Then, a network effects framework is developed in order to consolidate innovative insights into valuations in software markets.[6] But the framework has two functions, a descriptive and a prescriptive one (Grochla 1978). In addition to the descriptive identification of network effects in the context of valuations in software markets, the framework can also be used to derive descriptive investment and strategy guidelines for managers and financial sponsors. As the research fulfills both functions it serves as a heuristic tool that allows one to formulate and to solve respective problems in valuations in

[4] Please confer part II.

[5] Please confer chapter 7.1.2.

[6] Please confer chapter 8.

software markets (Kirsch 1984). In a next step, another content-analytic approach is applied in the complex networks analysis of hypotheses. This has the goal to challenge the most relevant hypothesis related to the properties, topologies and dynamics of complex customer networks in software markets.[7] Finally, the findings are integrated in another content-analytic research approach in a complex networks valuation framework for valuation in software markets based on the previously developed networks effects framework.[8]

1.3 Scale and Scope of the Research

The subsequent research seeks to identify the relevance of network effects and complex networks of valuations in software markets. However it must be noted that the existing research is vast. In order to investigate such complex interdisciplinary phenomena it is necessary to limit the explanatory parameters. Hence, the focus is on the outlined research questions, which thereby determines the scale and scope of the subsequent investigations.

1. *Sources.* The literature review is based on primary and secondary sources of research. Academic research papers are preferred over other sources of research, but complementary books and websites are considered if they provide meaningful research contributions. The majority of the resources are published between January 1995 and February 2009.
2. *Research Object.* As companies operating in software markets are the primary research object of this book, a corporate view is adopted. Other possible perspectives, e.g. welfare theoretical or individual views, are outside the research focus. Although the insights can also be applied to other industries, the primary focus of the subsequent analysis is on software markets due to the defined research objectives.
3. *Research Focus.* It is not in the scope of this paper to explore all implementation issues of valuations as the focus is on modeling of software markets for the derivation of cash flows and their respective volatility. The book pursues a customer network-centric perspective on valuation in software markets. Hence, this paper seeks to formally integrate properties, topologies and dynamics of networks in a network-theoretical approach for corporate valuation in software markets. Such an integrated valuation model underlines the value of each customer due to network effects and the importance of properties, topologies and dynamics of customer networks in the collective product diffusion process for valuation in software markets.

---

[7] Please confer chapter 12.

[8] Please confer chapter 13.

1.4 Target Audience

While research on customer network-centric valuation in software markets can be relevant to a broad audience in research and practice, it is particularly interesting for managers, financial investors and those in academics who are interested in valuation of software companies.

1. *Software Market Management.* Managers of companies operating in software markets frequently have to make vital managerial decisions, e.g. to pursue internal or external growth opportunities, based on results of investment valuations. The following research is of primary interest for software market managers, as insights derived from network effects and complex networks may have the potential to enhance the understanding of software markets.
2. *Financial Sponsors.* A majority of financial investors, e.g. private equity investors or banks, rationalize investment decisions based on valuations. As the investigations of networks effects and complex networks are useful in the enhancement of the quality of valuation in software markets, the subsequent research could be also interesting for them.
3. *Financial Research.* The valuation of companies is at the heart of financial research. Hence, the outlined customer network-centric valuation approach based on insights derived from network economics and complex networks contributes to the financial research community by relaxing some limitations of the traditional literature.
4. *Network Research.* The outlined research approach is also a contribution to research on network economics and on complex networks as both are applied in order to value customer networks in software markets. Thereby, the theories underline their potential for further applications in the social sciences.

It is important to note that the outlined interdisciplinary research objectives and the diversity of the target groups determine the writing style of the subsequent research. Although this book is not written for a natural scientific audience, the pursued research approach is quantitative in nature. Hence, it is a continuous challenge throughout the book to balance simplifications of abstractions, where possible, and the required rigorousity of academic research.

1.5 Course of Analysis

In order to achieve the previously outlined research objectives based on the depicted methodological considerations the course of analysis is structured as follows:

Part 1: Background and Motivation. The introduction has a twofold purpose as it provides a contextual and structural overview. In the first step, the relevant background of the research is outlined. Based on this illustration of the research motivation, the objectives of the research are clarified. The rationale behind the chosen research approach is revealed in the section on the research design, before the scale and the scope of the research is delineated. Then, the setup of the research design is legitimized by a description of the target audience. The introduction concludes with an overview on the course of analysis.

Part 2: Valuation in Modern Software Markets. The second part is the theoretical framework of the subsequent investigations. First, important concepts and underlying hypotheses of traditional investment valuation are summarized, before the specific profile of the real options approach is highlighted. Then, the profile of modern software markets is depicted in the subsequent theoretical block. Hence, their history and recent trends are provided, before the most relevant characteristics of software markets are profiled. The section concludes with a summary of requirements that have to be considered in valuations due to the specific nature of software markets. All insights are consolidated in a reconsideration of valuations in software markets in order to determine an approach for valuations in software markets that is capable of accounting for network effects. As a result of this analysis, the real options approach is identified as a concept that is capable of capturing and quantifying intangible flexibilities of management in software markets, before its implementation barriers are evaluated. In the parametrization of the real option approach, volatility and price are identified as the most important but also most difficult input parameters. Due to their complex nature and their high dynamics, it is even more difficult to determine them in software markets. All insights on the reconsideration of valuations in software markets are summarized in the final section of this part, which is the basis for the design of the subsequent framework for valuations in software markets.

Part 3: Modeling Network Effects in Software Markets. In part three the role of network effects for valuations in software markets is investigated. As the profile revealed the central role of network effects in software markets, the research on network economics is reviewed. Based on a sound understanding of network effects and customer-equity valuation, research on adoption and diffusion models is revealed as such approaches allow one to capture and to quantify network effects. A central research contribution in this context is the application of the Markov matrix analysis to investigate the diffusion dynamics of products in an analytic Markov Matrix Diffusion Model. All findings of the research are integrated in a network effects framework for valuation in software markets. This is based on numerical simulations of product diffusions in software markets that capture valuable characteristics and thereby allow for a disaggregated and more realistic corporate valuation. For this purpose, an overview of the framework is provided before the individual phases are described in detail. The part concludes with a reconsideration of the designed network effects framework that outlines the role of properties, topologies and dynamics of customer networks in software markets.

Part 4: Modeling Customer Networks from a Complex Networks Perspective. Due to the limitations of the network effects framework with respect to properties, topologies and dynamics of networks, the analysis is extended in the fourth part to complex networks theory. First, the relevant complex networks research

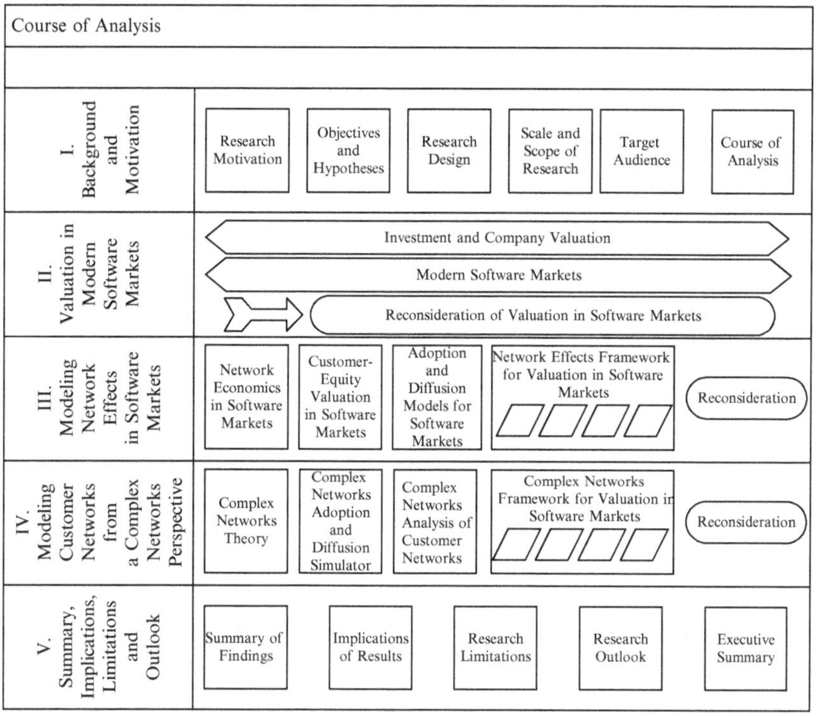

Fig. 1.1 Course of analysis
Source: Author

is summarized in order to derive relevant hypothesis that are investigated in the subsequent chapter. Then, a complex networks adoption and diffusion simulator is developed in order to investigate the hypotheses on customer networks from a complex networks perspective. The respective complex networks insights are integrated into a complex networks framework for valuations in software markets. Finally, the extended framework is challenged in order to derive the limitations of this customer network-centric valuation approach based on a complex networks adoption and diffusion simulator.

Part 5: Summary, Implications, Limitations and Outlook. In the last part of the book, a summary of findings and implications is provided, before an outlook on open research questions is composed. First, the main findings of the research are summarized, before its implications are derived. These are clustered with respect to the respective target group. Finally, the limitations of the research are discussed based on which further research opportunities are identified. An executive summary of the research and its main findings concludes the book.

Part II
Valuation in Modern Software Markets

This part is the theoretical framework for valuation in software markets based on which subsequent investigations are conducted. After a brief summary of the history of investment valuation, relevant classical and innovative financial concepts are outlined. First, conventional tools such as the asset value approach, the market value approach and the discounted cash flow approaches are depicted as well as the venture capital approach. Then, the more dynamic real options approach and the underlying option pricing theory are summarized. After this financial review, modern software markets are profiled. The principles and fundamental trends are sketched, before the most relevant characteristics of software markets are described. Both streams of research are combined in a reconsideration of valuation in software markets. The insights derived from this assessment determine the requirements for the design of the valuation frameworks in the subsequent parts of this book.

Chapter 2
Investment and Company Valuation

"Res tantum valet quantum vendi potest - The value is determined in the market."
Granger and Morgenstern (1970)

The most relevant concepts of the valuation literature are reviewed in this first chapter of the theoretical framework. First, the background of the relevant conventional valuation approaches is depicted, before the more innovative Real Options Valuation approach and the underlying Option Pricing Theory are summarized. The insights in this chapter provide the financial background for the subsequent design of a framework for valuations in software markets.

2.1 Principles of Investment Valuation

The review of research on *investment valuation* reveals the vital importance of recent research contributions. Capital budgeting research emanated originally as an individual stream of research focusing on economic resource allocation with the goal to determine the value of investment projects or assets.[1] Since the 1960s valuation tools in corporate strategy have flourished, due to the large emphasis on rational planning. A predominant strategic paradigm stated that accurate valuation and decisions about financial commitments are crucial for shareholder value creation and the survival of companies (Trigeorgis 1996). During the 1970s and 1980s the focus of financial research was on decentralized static investment projects. At the same time,

[1] The value of an asset is frequently defined as the sum of the subjective utility provided to its owner (Moxter 1991). While the neoclassical theory assumes that the price and the value of an asset are identical, more recent research on behavioral finance indicates that the two can differ (Shleifer 2000). In this context it is also important to note that motivations to conduct valuations are diverse and influence the outcome (Kühnemann 1985; Born 1995; Koller et al. 2005). Analogous to assets, the value of a company is defined as the total utility of a portfolio of investment projects. The company is interpreted as a set of temporary production functions (Busse von Colbe and Coenenberg 1992). Please note that the primary focus of the subsequent investigations is on asset valuation of customer networks in the context of company valuation as defined in Sect. 1.3. Hence, the terms customer network-centric valuation of companies operating in software markets and valuation are used interchangeably.

A. Kemper, *Valuation of Network Effects in Software Markets*,
Contributions to Management Science, DOI 10.1007/978-3-7908-2367-7_2,

the focus of the research was on valuation of stand-alone projects assuming passive management and certainty (Bamberg and Coenenberg 2004; Laux 2005). The methods developed until the late 1970's were not capable to account for the strategic value of the flexibility to alter plans (Trigeorgis 1996; Copeland and Antikarov 2001; Koller et al. 2005). Later financial research extended the set of applications to valuations under uncertainty. However the prevalent techniques were not capable of capturing all important aspects. Hence, the extension incurred frequently biased results and management decisions were based on managerial charisma. Consequently, managers made intuitive decisions in favor of strategic investments, based on the claim that the investment analysis does not account for all inherent flexibilities of risky projects.[2] This development caused a crisis in the research on valuations, in which it was difficult to identify market values due to a large theory-practice gap (Arnold and Hantzopoulos 2000; Boer 2002). While some academics argue that this gap is an intrinsic problem of financial research, others insist that the future is unpredictable, but suggest this should not be a preventive obstacle for research on valuations.

The resolution of the crisis was an article on a closed-form equation for financial option pricing which was published in 1973 (Black and Scholes 1973). This concept derived a theoretical price for all financial options and initiated a boost in the trading of derivatives. But the increased option thinking also affected other research areas. In 1977, Myers recognized that many projects handled by companies can be interpreted as *real options*. In the following decades these real options became one of the most promising valuation approaches and a large research area. During the 1970s and 1980s the application of dynamic investment approaches considering the reaction potential to uncertainties was gradually promoted to other areas (Trigeorgis 1996, p. 2f). Today, Real Options Valuation is successfully applied in many areas, such as the valuation of natural resources, electricity generation, and research and development investments (Coy 1999). Some models are even designed specifically for valuations in software markets (Schwartz and Moon 2000). Promising aspects of this concept are the consideration of the flexibility and the link between capital budgeting and corporate strategy. In theory, real option valuation looks like the perfect tool for managers to use as it provides more accurate values and normative investment decisions, but a review of its reach reveals that in practice it is not very popular (Graham and Harvey 2001; Koller et al. 2005). In this book some of the underlying problems are investigated and resolved.

2.2 Traditional Investment Valuation

Traditional corporate financial literature provides a variety of methods for the valuation of a company, which vary with respect to the required input data and the resulting level of detail (Busse von Colbe 1957; Ballwieser 1987; Moxter 1991;

[2] Risk is the possibility of an either favorable or unfavorable deviation from an expected value that is quantified by probabilities (Mikus 2001).

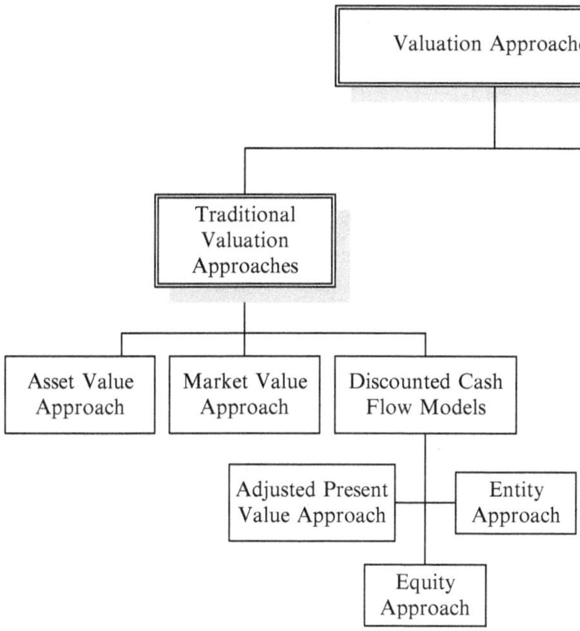

Fig. 2.1 Overview of relevant Valuation Approaches
Source: Author

Copeland et al. 1996; Ross et al. 1996; Brealey and Myers 1996; Koller et al. 2005). But due to the specific market characteristics only a few approaches are suitable for valuations in software markets. Figure 2.1 summarizes the most relevant models.

2.2.1 Asset Value Approach

The *asset value approach* states that the value of a company is equal to the sum of its assets valued from the perspective of either a going-concern or liquidation. If the going-concern of the company is assumed, the value of the company is equal to the sum of all operating assets plus the proceedings of all non-operating assets net of liabilities. In contrast, the liquidation value is the sum of all sales prices minus the liabilities and liquidation costs (Damodaran 1996). It is frequently used in a comparison with other valuations as it represents the minimum value of a company.

2.2.2 Market Value Approach

According to the *market value approach*, the values of assets and companies are derived based on proxies valued in financial markets (Born 1995). As there are a variety of proxies, the Similar Public Company Method (Gooch and Grabowski 1976), the Recent Acquisition Method, and the Initial Public Offering Method are each distinguished from one another (Bateman 1971). The respective multiples and ratios of traded companies are calculated based on accounting data, e.g. turnover or EBITDA margin, and allow a quick intra-industrial benchmarking analysis. All methods assume efficient capital markets and the Law of one Price which states that comparable assets are supposed to have the same prices as arbitrage would be otherwise possible (Fama 1970). Thereby, risk is implicitly treated as the risk premium is contained in the multiples of the proxy and frequently adjusted with an additional risk premium or an investment specific discount. In practice, market value approaches are very popular. Empirical studies comparing the popularity of valuation tools reveal that in nearly 73% of all international financial transactions marked based approaches are applied, second only to Discounted Cash Flow models which have a predominant role in valuation with a popularity of 95% (Peemöller and Kunowski 2002). Nevertheless, the approaches do not provide a true and fair view as the approach does not adequately incorporate the idiosyncratic risk profile of the valuation target.

2.2.3 Discounted Cash Flow Models

The *Discounted Cash Flow* (DCF) paradigm is the most popular instrument for estimating the value of both projects and of companies. It states that the respective value is the cumulated present value of expected predicted cash flows and a normalized terminal cash flow discounted by a respective risk-adjusted discount rate (Brealey and Myers 1996; Koller et al. 2005). In this concept the risk-adjusted discount factor represents the cost of capital which can be derived by capital market models such as the *Capital Asset Pricing Model* (CAPM) or the *Arbitrage Pricing Model* (APM).[3] Another critical aspect of the DCF is the calculation of the terminal value as it can account for more than 80% of the total corporate value and is even higher for software companies (Brealey and Myers 1996). Literature provides a variety of approaches for valuing the terminal value, such as the Gordon Growth model, the Multi-Stage Growth model or the convergence model (Ross et al. 1996). In essence, the approaches assume that the terminal value can be derived by a perpetuity growth rate, which is frequently approximated by an industry average (Gordon 1959; Koller et al. 2005). The various DCF approaches can be distinguished into Entity-, Equity-,

[3] The opportunity costs of capital are the sum of interest for equity and debt financing. The most influential cost of capital concepts are the CAPM and the APM (Sharpe 1964; Lintner 1965; Mossin 1966; Ross 1977). While the original model was published as Arbitrage Pricing Theory with a focus on securities, in the following the broader term Arbitrage Pricing Model is used which also comprises publications on the pricing of derivatives. Please consider (Ross et al. 1996) for further information.

and Adjusted Present Value approach based on varying assumptions with respect to the cash flows, the capital structure, and the underlying taxation system.[4]

2.2.3.1 Entity Value Approach

In the first step of the entity approach, the total enterprise value is computed based on an all equity financing fiction, before the value of debt is deducted in order to determine the residual value of the equity. The total enterprise value V is equal to the sum of all gross cash flows discounted by the *Weighted Average Cost of Capital* (WACC), such that

$$V = \sum_{t=1}^{T} \frac{E(CF_t)}{(1 + WACC)^t}, \tag{2.1}$$

with the investment time horizon T and the cash flows CF_t at time t (Copeland et al. 1996; Koller et al. 2005). In order to derive the equity value EV, the market value of interest bearing net debt DV is deducted

$$EV = \sum_{t=1}^{T} \frac{E(CF_t)}{(1 + WACC)^t} - DV. \tag{2.2}$$

The additional value of a tax shield is reflected by the term 1− corporate tax rate s in the calculation of the WACC, which is defined as

$$WACC = E(R_D)(1 - s)\frac{D}{E + D} + E(R_E)\frac{E}{E + D}, \tag{2.3}$$

where expected returns of the equity and debt market $E(R_D)$ and $E(R_E)$ have to be derived by the CAPM or the APM. The computation of the WACC implies a circular problem as the determination of the WACC requires the enterprise value ($V = E + D$) which, in turn, should be determined with the help of the WACC. This circular problem can be resolved by assuming a target capital structure or iterating the capital structure (Ross et al. 1996).

2.2.3.2 Equity Value Approach

The equity value approach directly determines the net value of the equity (Ross et al. 1996; Koller et al. 2005). Net cash flows are the financial flows to the shareholders that can be withdrawn from the company. Accordingly, the present value is calculated by discounting available cash flows to the holders of equity capital,

[4] Please consider (Brealey and Myers 1996) for further details on DCF models.

after allowing for cost of servicing debt capital. Formally, the equity value of the company is

$$V_E = \sum_{t=1}^{T} \frac{E(CF_t^E)}{(1+r)^t}, \tag{2.4}$$

with the company value V_E, the time horizon T, and the opportunity costs of capital r. The cash flows CF_t^E at time t are defined as

$$(E_t - A_t) - ((E_t - A_t - r_{FK} \cdot FK_{t-1}) \cdot s - r_{FK} \cdot FK_{t-1}) - (FK_{t-1} - FK_t) - C_t, \tag{2.5}$$

with E_t as inflows and A_t as outflows at time t, C_t as capital expenditures, FK_t as the market value of debt at time t, with r_{FK} as the cost of debt and s as the corporate tax rate. The cost of equity can be determined, e.g. with the help of the CAPM as $r_{EK} = r_f + [E(r_m) - r_f] \cdot \beta$, with $E[r_m]$ as the expected return of the market portfolio, the risk free rate r_f and the beta factor β.

2.2.3.3 Adjusted Present Value Approach

A third DCF approach is the *Adjusted Present Value* (APV) model, in which the value of the company is deconstructed into several elements driven by the purpose of isolating the value of the tax shield. First, it is determined which cash flows of the company can be withdrawn based on full equity financing and discounted at the equity financing costs in order to determine the operative value of an unlevered company. In the final step, the tax benefits resulting from the tax shield are added and the respective adjusted present value is derived (Ross et al. 1996).

2.3 Real Option Valuation

Software companies operate in a dynamic and competitive environment with a high exposure to uncertainty.[5] In such situations real options provide their owner in analogy to financial options with the right to exchange the cash flow of an underlying against the value of an exercise price (Koller et al. 2005). In other words, there exist additional operative or strategic flexibilities to pursue profitable opportunities which increase the value of an investment project or of a company. Such asymmetrical pay-off structures, resulting from managerial flexibilities, can be interpreted as valuable real options. Further examples of real options derive from flexibilities to defer, to change or to terminate an investment (Trigeorgis 1996).[6]

---

[5] Please confer Sect. 3.3.

[6] Please confer Sect. 2.3.3 for a typology of real options.

2.3.1 Option-Pricing Theory

Real option valuation (ROV) stands on the shoulders of the *Option Pricing Theory* (OPT).[7] Accordingly, an option is defined as a financial contract that provides its owner with the right but not the obligation to exchange an asset or a financial contract against another at a given price at the expiration date (Black and Scholes 1973; Margrabe 1978; Fischer 1978; Stulz 1982).[8] All options share three essential features, namely

1. flexibility,
2. uncertainty, and
3. irreversibility.[9]

If the prerequisites are fulfilled, the option pricing theory can be applied in order to value a spectrum of options, ranging from simple plain vanilla options to more complex options, such as compound options, options with a stochastic underlying, with a stochastic exercise price, with different exercise times, or with varying convenience yields (Hull 1989). All types of options have in common that the pricing is based on risk neutral valuation by assuming a perfect and arbitrage-free financial market in which a replicating portfolio of traded securities is constructed in order to mimic the payoffs of the option (Ross 1977). Cash flows are discounted by a certainty-equivalent growth rate representing a risk premium that would be appropriate in a market risk-neutral equilibrium in order to transform the expected cash flows with equivalent martingale measures into objective probabilities. The *Law of One Price* and *dynamic tracking* allow one to assume that the value of an option is equal to the value of the twin portfolio as there would otherwise be arbitrage opportunities (Merton 1976). Hence, the goal is to replicate the cash flows of the investment by a portfolio of twin-securities with returns perfectly correlated to the underlying.[10] In perfect capital markets a synthetic hedge can be designed by combining the twin securities and risk-free bonds such that the fixed final pay-offs of the correlated security and the cash flows discounted at the risk-free rate match the value of the option (Trigeorgis 1996; Koller et al. 2005). In other words, if investments have the same payoffs, they are supposed to have the same prices. Consequently, the value of the option can be determined by replicating the investment and value the replication, based on the assumption that in perfect capital markets any arbitrage opportunity would result in a revaluation of the asset or the derivative.

---

[7] Please confer (Wilmott et al. 1995) for an extensive overview on option pricing theory.

[8] More generally, the option embeds the right to purchase or to sell an underlying at a predetermined exercise price on (European) or before (American) a predetermined date. Values of options can stem from two different sources, the *intrinsic value*, which is equal to the price differential between the underlying and the exercise price, and from the *time value* until expiration (Myers 1977).

[9] Please confer (Hull 1989; Trigeorgis 1996; Neftci 1996) for further information.

[10] The portfolio is also called tracking portfolio (Amram and Kulatilaka 2000).

2.3.2 Real Options Analogy

Software companies operate in a risky and competitive environment. They exhibit an asymmetric payoff structure due to managerial flexibilities, which can be interpreted in an analogy to financial options as real options (Myers 1977). Accordingly the equity holders are considered to own options on the total asset value less the current debt burden. Thereby, the real options quantify the additional operative or strategic flexibilities that increase the value of an investment project or a company, such as flexibilities to defer, to change or to terminate investments (Trigeorgis 1996; Koller et al. 2005).[11] While the approach was originally designed to value single real options corresponding with single investment projects, it can be extend to aggregated company valuation by interpreting the company as a portfolio of options.[12] With respect to valuation in software markets real options allow managers and financiers to capture the value of customer networks in software markets as growth options if the required option parameters are available. Accordingly, a call option on the cash flows of the company is purchased for a relatively small investment into the customer network of software companies, which can be obtained once the customer network reaches a critical mass. In other words, the overall value is decomposed in a risk-less and a risky component. While the passive component can frequently be determined with a static DCF calculation, the optional value of the growth option is determined by a real option valuation which can comprise one or multiple interacting options. In sum, the overall value is given by

$$V_U = V_S + V_D, \tag{2.6}$$

with a static value V_S and a dynamic value V_D. Although some simplifying assumptions may be required in order to derive the parameters for the valuation, the additional investigations of the managerial flexibilities based on ROV can significantly increase the quality of the valuation. This is particularly relevant if the investment is subject to high managerial flexibilities and a high degree of uncertainty. Both conditions are frequently met in software markets.

2.3.3 Typology of Real Options

Within the field of research there are a variety of classifications for real options (Trigeorgis 1996, p. 2f) (Lander and Pinches 1998, p. 540). The most suitable typology for valuation in software markets is based on the varying flexibilities of the respective underlying (Hommel and Pritsch 1999, p. 125).

[11] Please note that the value of a real option is zero once it is exercised, but it can not have a negative value as it is a right and not a binding obligation (Hommel and Müller 1999).

[12] Please note the interdependency of multiple real options.

1. *Deferral Options. Deferral options* are options on cash flows in exchange for the initial investment providing a flexibility to extend the deadline. Therefore, they are also coined *options to wait*. As far as the modeling is concerned, the option to wait can be interpreted as a call option on future cash flows in exchange for an adequate option premium.
2. *Liquidation Options. Liquidation Options* are put options that allow one to terminate an investment earlier than initially expected. It is also termed *option to abandon* or *exit option*.
3. *Shut-Down Options. Shut-down options* allow one to interrupt the production for a certain time. They are also called *options to shut down and restart*.
4. *Continuation Options. Continuation options* results from the multiple stages of an investment financing. This class of options is also labeled *options to stage investments*, which can be modeled as a compound option.
5. *Scale-Up and Scale-Down Options. Scale-up and scale-down options* provide the owner with the right to extend or contract the production if market conditions require an adjustment of the production volume. They are also termed *options to expand or to contract* business.
6. *Switching Options. Switching options* contain the flexibility to choose between various input factors and are also termed *exchange options*.
7. *Innovation Options. Innovation options* provide the flexibility to choose between various production processes. They enable owners to benefit from follow-up projects and have the payoff profile of a *compound option*.

The outlined options can be grouped into the following three clusters (Fig. 2.2).

(a) *Learning Options*. These options provide their owner with the flexibility to delay an investment until more information is accessible. The delay is particularly valuable if large irreversible investments are investigated. Examples are the option to wait and the option to stage the investment.
(b) *Growth Options*. Growth options are the second class of options that allow their owner to benefit the scale of operations under positive environmental conditions. Examples are the options to innovate and the options to expand as they allow the exploitation of further growth potentials. This class of options is frequently the most valuable type of real option in the valuation of turnarounds.
(c) *Insurance Options*. Insurance options comprise options to alter scale, switching options, options to abandon investment and the option to stage an investment. These options allow a reaction to unfavorable market developments. They allow one to reduce the downside risk and the overall volatility of the project.

Besides the functional distinction, other characteristics to classify options are their interdependencies and their time horizon. Independent options are valued on a stand-alone basis, while *compound options* are interdependent (Geske 1978).

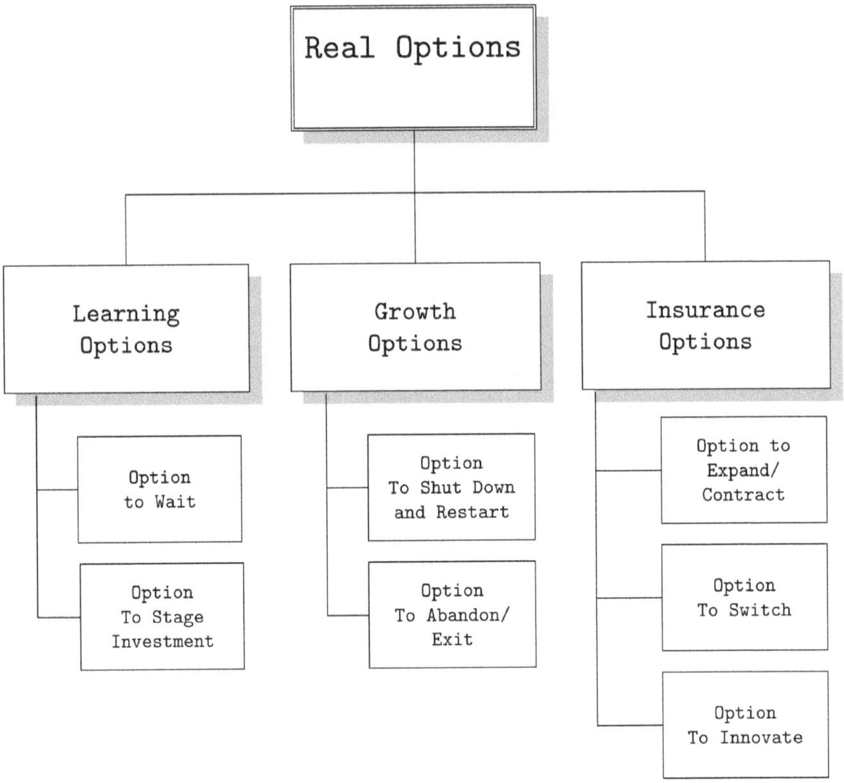

Fig. 2.2 Real Option Typology
Source: (Hommel 1999)

2.3.4 Real Options Management Process

While there are various frameworks for structuring Real Option valuation, the integrated Real Options Management Process of (Hommel 1999) consists of three interdependent steps which are depicted in Fig. 2.3 (Hommel 1999).

1. *Identify Real Options*. In the first step, the relevant sources of uncertainty and respective flexibilities have to be identified. At the same time it is important to consider the limitations of the option analogy in order to determine the scale and scope of the valuation model.
2. *Valuation of Real Options*. After the identification it is necessary to quantify and to value options and their interactions. Therefore, a suitable option valuation approach has to be identified before the respective parameters of the approach are derived. If multiple real options are involved, the interaction of real options has to be considered. Later it is necessary to assure the rationality of the valuation, which can be achieved by stress testing the underlying assumptions.

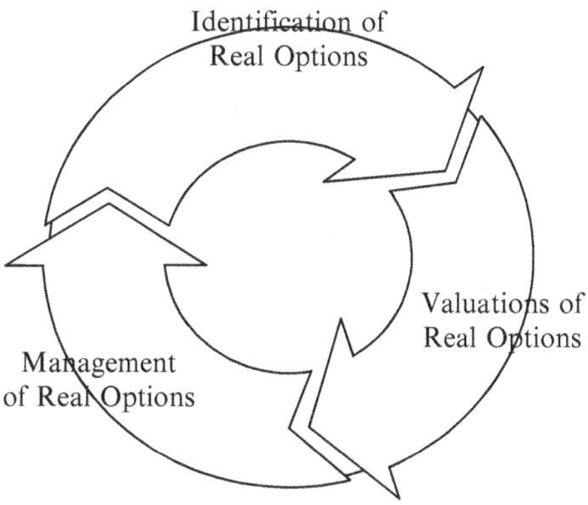

Fig. 2.3 Real Options Management Process
Source: (Hommel 1999)

The resulting alternatives have to be compared in order to determine the most promising strategy.

3. *Management of Real Options*. Once the options are valued, management should implement strategies that increase the overall value of the company. This can be achieved by integrating option pricing strategies into strategic management. Identified real options have to be tracked. But at the same time it is important to continue the identification process of new Real Options. Proactive management of Real Options should be pursued by influencing the value drivers of real options in order to gradually increase the value of the identified options.

Alternatively, Cropland suggests a four-step process model that could be adjusted to valuation in software markets (Copeland et al. 1996, p. 417). While the first two phases have a preparatory character, the last two comprise the quantification. On a finer level of detail, the interdependent phases can be described as follows.

1. *Design of a DCF Model*. In the first phase of the model the competition, the price and other market parameters are investigated in order to assess the overall importance of relevant real options.
2. *Risk Assessment*. The base case of the DCF valuation is evaluated with respect to the main sources of uncertainty such as inside technological or outside economic risks and the respective managerial flexibilities. In this context major decisions and milestones have to be identified, which can be supported by the outlined classification scheme of relevant Real Options.[13] Once the options are identified, their development of multiple options can be modeled based on event-based decision trees, while single options can be represented by simple time-line graphs (Copeland and Antikarov 2001, p. 418).

[13] Please confer Sect. 2.3.3.

3. *Simulation Phase.* After a determination of the options, they have to be quantified, which can be achieved by either analytical or numerical approaches.[14] Therefore, the gross value of the company is determined for various points in time with respect to the identified uncertainties and decisions, while interactions of options are ignored. At the same time, competition can be incorporated into the model by adjusting the dividends.
4. *Interactions among Real Options.* In the final phase, the interactions among real options are considered. The can have nonlinear sequential or simultaneous impacts on each other that have to be taken into consideration.[15]

Although this process model contains important issues that are relevant to valuations in software markets, some adjustments are required in order to account for the industry specific characteristics, which will be investigated in the subsequent chapter on modern software markets.[16]

2.3.5 Real Options in Practice

Empirical studies investigating the preference of practitioners with respect to investment valuation tools revealed that both the acceptance of financial instruments is different in various countries, and that for a variety of reasons it always requires some time (Peemoeller et al. 1994; Trigeorgis 1996; Copeland et al. 1996; Leslie and Michaels 1997; Koch 2000). While the Internal Rate of Return was the predominant valuation tool in Germany during the 1970s, in the 1980s, the DCF approaches reached a similar popularity. Research indicates that real options have become increasingly important, but remain a complementary tool to the traditional valuation tool kit. Driven by the interest to explain high valuations despite negative cash flows, research on Real Options experienced a boost during the years of the Internet-hype. But this interest diminished after the stock market crash. Nevertheless, real options analysis is applied nowadays in several industries, such as construction, resource intensive industries, biotechnology companies, media companies and strategic consultancies. Empirical research on the popularity of valuation approaches illustrates that the importance of valuation instruments increased over time, and most studies find a low but increasing percentage of advanced valuation techniques (Ho and Pike 1991).

Investigations concerning the quality of the Market Value and the DCF approaches come to contradictory findings, as a comparison of the valuation results with the respective stock price reveals that frequently very rigid assumptions diminish the explanatory potential of valuation models (Copeland et al. 1996; Damodaran 2001).

---

[14] Please confer Chap. 4 for further reference.

[15] Please confer the sensitivity analysis in Sect. 8.5 for details on possibilities to account for such interdependencies.

[16] Please confer Chap. 3.

The oversight of real options is a reasonable explanation for this low performance. An analysis of German NEMAX 50 companies compared a passive DCF value with the stock price and found that only 16% of the value was explained by the DCF values, while the residual 84% were attributed to options (Rohjahn and Berner 2002). In a similar study, a passive DCF value accounts for only 6% of the overall value while the residual optional value is 94% (Stemmann and Treptow 2001). Despite the great importance of optional values, a survey reveals that 84% of the interviewed companies do not apply real options, while 8% do not even see a necessity for a real option valuation as they are convinced that intuitive risk adjustments, decision-tree analyses, etc. are sufficient tools. (Vollrath 2001) The low popularity is confirmed by an additional study, stating that at that time only 22% of the German DAX, NEMAX and consulting companies explored real option valuation (Peemöller et al. 2002). The main reasons for this low popularity are the complexity of the approach, lack of experience with more complex approaches, and nearly a third of the interviewed managers are convinced that other valuation tools lead to similar results. Therefore, despite the aforementioned contributions, it is very unlikely that the real options approach will replace the predominant DCF approach in the near future. It will allow rather a complementary quantification of managerial flexibilities, as the subsequent discussion illustrates.

2.3.6 Reconsideration of Real Options

Real option approaches relax some of the limitations of traditional approaches, as they allow one to analyze investment decisions under uncertainty, and account for managerial flexibilities (Trigeorgis 1996). A central advantage is that neither subjective biased probabilities nor subjective risk preferences are required. Instead, values are derived from the quasi-objective capital market based on the risk-neutral valuation principle. Therefore, some researchers consider real options to be the new standard valuation tool in corporate finance, while other financial researchers are not convinced that real options are suitable to value real world flexibilities as they are constrained by the following six most relevant limitations (Kester 1984; Schluechtermann 1996; Kruschwitz 2005).

1. *Intuition.* The Real Option Approach is not very intuitive as it requires knowledge of option theory. But as derivatives are part of many courses in modern management education, the popularity of the option pricing theory is likely to increase over time.
2. *Competition and Competitive Interactions.* Competition can dilute the value of real options. While all financial options are proprietary to their owners, as they guarantee an exclusive right, not all real options are proprietary. Some real options can contain legal rights, such as patents etc., while other types of real options are collective property, such as first mover advantage growth options. Such options are shared by all competitors and are difficult to protect, which can imply a dilution of their values.

3. *Illiquidity and Transaction Costs*. Financial options are fungible. They are traded in efficient financial market at minimal transaction costs, whereas real options are rather illiquid. Particularly real world capital budgeting projects are not traded continuously. Therefore, is doubtedful if a replicating portfolio can be constructed. Alternative proxies have to be used which are effected by tracking errors. In addition, real options depend frequently on other options or assets. Such a collective portfolio is even less fungible as it requires to align the interest of multiple owners.
4. *Strategic Interactions*. The value of financial options vitally depends on the value of the underlying. Similarly, the value of simple real options depends on the development of their underlying, while the value of more complex real options can interact with other flexibilities, such as further discretionary investment opportunities. Such options provide options on options, which are also termed *compound options*. They are more complex to evaluate as multiple options are non-additive, the complexity of the approach increases with an increasing number of relevant options, whereas the marginal contributions of additional options are still positive but decreasing (Schwartz and Moon 2000). Since six real options frequently cover over 90% of the total optional value, the focus should be on the six most important sources of flexibility and uncertainty (Trigeorgis 1996).
5. *Computational Complexity and Information Requirements*. Depending on the implemented Option Pricing Model, the computational requirements of the real options approach are more challenging than those of other approaches. It is possible to observe that over time the computational capacities of standard computers steadily increase. Therefore, it is becoming increasingly easier to implement even challenging option pricing models with home computers. But the implementation of a real options approach always requires a variety of detailed data which is frequently unavailable.
6. *Standard Stochastic Processes*. Empirical studies testing the Moon and Schwartz model come to opposing conclusions with respect to the precision of the utilized standard stochastic processes. While Schwartz and Moon test the model in a clinical study and reach a positive conclusion, other researchers come to different conclusions. Accordingly, empirical results reveal that the Schwartz and Moon model is an important research contribution but should be improved in order to reach a higher reliability (Keiber et al. 2002).

Some of the outlined restrictions can be resolved by advanced valuation models. But such advanced models cause a dilemma between their accuracy and the respective costs (Meise 1998; Hommel and Müller 1999; Kühn et al. 2000). As the quality of input parameters determines the quality of the analysis, prohibitive information costs can prevent the application of the real options approach. In turn, it is possible that the incremental gain of information outweighs the required additional effort. Based on the insight that the Real Options approach has the potential to contribute to valuations in software markets, the following research is dedicated to identifying implementation barriers and room for improvement for valuation in software markets. For this reason, the background of software markets is studied on a finer level of detail in the subsequent chapter.

Chapter 3
Modern Software Markets

> "If General Motors had kept up with technology like the computer industry, we would all be driving twenty-five dollar cars that got 1.000 miles to the gallon."
>
> Bill Gates, at the computer expo (COMDEX) 1998

The focus of the second chapter of the theoretical framework is to profile modern software markets. First, a brief review of the history is presented which illustrates the background of the main research objects, before fundamental trends in software markets are summarized. In the next step, essential characteristics are identified and outlined. As the profile indicates that software markets are governed by network effects, their impact on valuation in software markets will be reconsidered more closely in Chap. 4.

3.1 Principles of Software Markets

A review of research on *software markets* reveals that the relationship between customers and software changed significantly over time.[1] In this book software markets are defined as markets of intangible software programs, services, and applications. Researchers agree that the development of software markets can be separated into five phases of commercialization. These illustrate the changing importance of customer networks in software markets (Hoch et al. 1999):

1. *Integrated IT Projects*. The development of software markets dates back to the split-up of the computer market into a hardware and a software segment. Until the 1960s the development and production of hardware and software were closely related, as the software was an integral part of the hardware. Governmental

[1] *Software companies* operating in software markets are the main research objects of this study. They are defined as hardware independent firms focusing on research, development, distribution and maintenance of software programs (Gerhardt 1992). *Software programs*, in turn, allow the execution of information processing operations in order to create, transform and analyze data and information. The broader term *computer software* describes a collection of programs, procedures and documentation that perform operations on a computer system (Baldi 1998).

A. Kemper, *Valuation of Network Effects in Software Markets*,
Contributions to Management Science, DOI 10.1007/978-3-7908-2367-7_3,

organizations and large industrial corporations worked on highly specific and integrated software and hardware projects, before the second phase when independent software programs were developed. Initially, customer-made software solutions for individual customers were designed for the US government. *SAGE* was an air defense software project with a budget of Eight Billion US dollars. 700 out of the 1,200 governmental programmers worked on the project which was developed between 1949 and 1962. *SABRE* was another governmental software project for flight reservation service system which was developed between 1954 and 1964. After the completion of these projects many of the programmers founded independent companies and launched the emergence of a new industry. In 1955, before the term software even existed, two IBM programmers founded the Computer Usage Company and provided, for the first time, independent hardware computer services and software. Other companies emerged such as the Computer Sciences Corporation, Applied Data Research, Management Science America, etc. many of whom are still in business today. With the increasing number of computers, the first half of the 1960s is characterized by huge growth rates of the new industry due to a strong demand for software solutions. In 1965 the US software market consisted of 45 large software companies and 2,800 small-sized firms, while the European software markets emerged with a time lag in the 1950s and 1960s. While some of the initial software pioneers are still competing in the contemporary software markets, others disappeared. Valuations in software markets have already been a topic of vital importance the individual software projects had to be valued. While the purpose of valuations was initially an analysis of capital budgeting decisions their role increased with the developments as the second phase began.
2. *Individual Software Products*. The second phase spans from 1959 until 1969 and is characterized by the emergence of smaller and more specialized software products as a direct result of larger software projects. While many members of the computer industry assumed that the development of software was not profitable, a few believed in a software market for standardized solutions. In 1964, ADR developed Autoflow, a flow chart program, which was the first software product with a marketing strategy. It was sold to several thousand users and was protected by the first patent on a software program. Another commercial success was the program Mark IV from Informatics with a sales volume of more than USD 100 mn from the first software licence royalties. Hence, the software pioneers of the 1960s built the foundation for the modern software market architecture, influencing the design process, pricing, maintenance and copyright protection of software. In this phase, valuation in software markets became crucially important as it was necessary to determine values for companies in the context of corporate transactions, however, the support of the financial theory was rather limited.[2]
3. *Separation of Hardware and Software*. The design of independent office solutions for a broader customer base marks the third step in the development of

[2] Please confer Sect. 2.2.

software markets. A broad variety of software providers emerged between 1969 and 1981 due to the split-up of the computer market into two different segments, i.e. hardware and software (Neugebauer et al. 1983). In 1964, IBM launched the IBM/360 which was the first computer with modular Hardware components. The product was installed on approximately 50,000 computers, representing a market share of more than 80%. This success allowed a sales volume of USD 16 bn and profits of USD 6 bn. Increasing competition forced the industry to specialize on core competencies and in 1969 IBM decided to separate the sale of software and maintenance services from hardware components. This step revealed the value of software and created a market for software services and products. It marks the emergence of software markets. In the following, alliances of hardware providers and external software companies assured the development of operating systems, office applications, and computer games, all of which, in turn, increased the sale of hardware. Initially the insurance industry fostered the development of independent software solutions, but other industries followed mainly due to the need for database applications. Funded by the venture capital industry, many start-ups emerged and provided products for emerging market niches. In 1972, the German software company Systems, Applications and Products (SAP) entered the software market with a standardized business process software. After eight years SAP had sales of USD 60 mn and supplied half of the largest industry companies in Germany. Another eight years later, SAP went public with a sales volume of USD 200 mn. Today, SAP ERP is a market leading product in the ERP segment. In the third phase other software players emerged, such as Computer Associates (1976), Oracle (1977), Baan (1982), Sybase (1984) and Peoplesoft (1987). In the 1980s and 1990s, the role of valuation in software markets was further increased as external growth by acquisition became increasingly popular.

4. *Software Mass-Markets*. In the fourth step, standardized products were offered to a global mass-market. While many companies concentrated on customized software solutions, the demand for mass software products steadily increased. At the same time, emerging standards and customer networks provided platforms for upcoming innovations. The rise of the PC required the production of software for the mass-market, which in turn required new marketing and distribution strategies. In 1977, Apple presented the Apple II and in 1979 the first spreadsheet software VisiCalc. Although it was a killer application, Apple was not able to establish a market standard due to its low penetration rate. Instead, in 1981 the IBM PC was introduced which marked the birth of the independent modern software market. Microsoft, founded in 1975 by William Gates and Paul Allen, developed the operating software MS DOS for the PC. Its successor program, Windows, became the predominant market standard and provided Microsoft with financial resources to strengthen its position in the application software markets. Based on the IBM platform, a variety of new software companies emerged such as Lotus (1982), Adobe (1983), Autodesk (1983), Intuit (1983) and Novell (1983).
5. *Internet*. Finally, the emergence of the Internet introduced a new era of software solutions for interacting customer networks. The rise of the Internet marked

an additional growth factor of software markets due to the universal network opportunities, such as electronic market places, offering business-to-business, business-to-consumer as well as consumer-to-consumer solutions. A variety of standards coexisted in various software market segments, but at the same time the Internet provided a source for large business opportunities as the growth of the software market outperformed many other industries. While the total market volume was approximately USD 0.3 bn in 1983, several years of double-digit growth rates lead to a market size of USD 1.2 bn in 1978, USD 3 bn in 1984 and USD 300 bn in 1997 (VDI 1983; Groehn 1999). Market entry barriers were comparatively low as programmers, developers and consultants offered services without significant financial investments. Despite this increase in market volume, a large proportion of software companies were driven out of the competitive markets. In particular the fall at the turn of the century induced a consolidation in which more than 70% of the turnaround activities failed. In this context valuation in software markets played a crucial role in this selection process, as the results influence the turnaround assessments of financial investors and the market for corporate control.

This review outlines the development of customer networks during the development of software markets. While software was sold initially as part of the hardware to some selected customers, it is nowadays sold to a global network of users. The most important fundamental trends in software markets are outlined in the following section.

3.2 Fundamental Trends in Software Markets

Software markets are characterized by changes. In addition to fundamental megatrends, there are also industry and sector specific microtrends that determine the development of software markets. The most relevant trends for valuation in software markets are depicted in the following section.[3]

1. *Standardization and Consolidation.* European software markets are very fragmented. Even large software companies have rarely a market share of more than 10–15%. (Citigroup 2006) But they are increasingly characterized by sector consolidation as companies want to improve the total costs of ownership by reducing customization and integration efforts through standardization of software vendors. Customer networks play a decisive role in this standardization and consolidation trend. Hence, the main benefactors are companies with large customer networks, whereas small vendors are expected to team up or sell themselves to larger competitors.

[3] Please confer (Naisbitt et al. 2001) for additional information on megatrends of software markets.

2. *Media Convergence*. IT, Telecommunication and Business services are converging (EITO 2006). This convergence is driven by two factors. First, the increased demand for outsourcing creates an area where IT and business services are competing for the same deals. Second, the cannibalization of fixed line revenues by Voice over IP is driving Telco vendors into the IT Service market. But the convergence between IT and Business services is fostered by outsourcing contracts and by IT vendors moving into traditional business services. A second convergence trend comprises IT and Telecommunication services, as operators tendering for total infrastructure outsourcing deals tend to subcontract such areas.
3. *Outsourcing and Offshoring*. The supply chain of software companies changes as an increasing part of the value chain is outsourced or offered with off-shore capabilities in India, China and the Philippines (Strahringer and Westner 2008). Such measures lower the costs per capita and increase the pressure on small local vendors. This trend is reinforced by innovative "*Software as a Service*(SaaS)" business models that allow companies to rent a software for a flat fee (Beinhauer et al. 2008). They combine the possibilities of outsourcing and renting software instead of buying it, which allows companies also to change their spending priorities.
4. *Service Oriented Architecture*. Software is expected to increase its weight of total IT investments. Empirical surveys reveal that software spending continues to be on the rise, while investments in hardware are either stable or even decreasing (Citigroup 2006). This trend is reinforced by *service oriented architectures* (SOA), that reduce the need for third-party developments and traditional middleware. The SOA paradigm changes the design of business applications, enabling the rapid composition of business solutions, e.g. the encapsulation of business logic and exposition as enterprise services. Applications are no monoliths. They consist of various modules. While object-orientated design created too small units to exchange them on a broad business scale, services are larger but also flexible. Small functional components are re-assembled quickly to develop new business solutions that meet changing business requirements based on generally agreed standards. Thereby, SOA allows corporates to change their spending priorities.
5. *Cloud Computing*. Cloud computing describes the trend of cloud-based, i.e. Internet-based, development and use of computer technology (Armbrust et al. 2009). Accordingly, dynamically scalable and often virtualised resources are provided as services via the Internet. A specific characteristic of cloud computing is that users do not have to care for the technology infrastructure that supports the requested services. Depending on the exact definition cloud computing incorporates concepts that satisfy the needs of customers via the Internet, e.g. *infrastructure as a service* (IaaS), *platform as a service* (PaaS), or *software as a service* (SaaS).
6. *Open Source Software*. The trend towards open source software is expected to free up some development and production budget of companies. In total, the net impact should lead to a reduction of more than 50% of overall software budgets for infrastructure tools (Citigroup 2006).

7. *Enterprise Resource Planning*. In the ERP sector, nearly all larger companies aim for a business process platform. While major financial services companies are ready to invest, retailers are focusing on innovation and are willing to modernize their ERP systems. It is important to note that there exist interfaces in ERP markets to exchange data among systems of various companies.

In addition, the following fundamental trends concerning various types of software can be differentiated.

1. *Application Solutions*. As flexible component-based applications improve the level of re-use, companies invest in IT in order to improve user productivity and enhancing process efficiency. Key growth areas are collaborative applications in management information systems, supply chain management, customer relationship management, human capital management and procurement. In parallel, the increasing demand for financial decision support software is another driver increasing the demand for application solutions.
2. *System Software*. The demand for system software is mainly driven by security software, storage software and content access software. Security enforcement tools such as Anti-virus protection fuel the growth as they are increasingly required by regulation. Complementary services are storage replication and data recovery.
3. *Operating Systems*. Microsoft had long a dominant position in the operating system market for personal computers. Recently, various developments threaten the hegemony of Microsoft. For one, software is directly embedded into semiconductor chips. This reduces the need for separate operating software. In addition, there is a trend toward low-priced terminal devices that connect directly with the Internet and operate with hardware independent software.

The selected trends illustrate that software markets are at a major inflexion point as the new business cycle implies fundamental changes. The changes in spending will undoubtedly have implications on the value chain in software markets. Particularly, the push towards SOA and the move from software vendors to increase the integration between enterprise applications and infrastructure software will cause restricted growth, price pressure and an accelerated commoditization. In addition, this trend will reinforce the importance of customer networks in software markets due to reinforcing network effects. Similar effects result from the convergence between IT and telecommunication services. All outlined trends are based on the underlying characteristics of software markets, which are reviewed in the next section.

3.3 Characteristics of Software Markets

The profile of software markets differs from that of other industries. With respect to valuation in software markets, the following selected characteristics are particularly relevant:

3.3.1 Short Product Life Cycles

Depending on the underlying nature of the software, most software markets are characterized by a short product life cycle (Levitt 1965; Day 1981; Box 1983). Commonly an iterative succession of four stages is assumed in which the ratio of commercial costs and sales changes. The size of the customer network and the respective turnover is the decisive criterion to distinguish various phases of the product life cycle as depicted in Figure 3.1.[4]

1. *Market Introduction Phase*. The first phase requires large investments in research and development, while initial sales volumes are rather low. Despite of only few competitors, other firms will observe the market development very closely. As it is necessary to reinforce the demand, a convincing marketing strategy has to attract customers. While a few early adopters are likely to join, the majority of customers prefers to observe the market development.
2. *Growth Phase*. The expansion phase is characterized by cost reductions due to increasing economies of scale in the increasing customer network. Hence, the sales volume increases significantly. This has a positive impact on the profitability and the financing of further expansion plans if the respective costs remain relatively stable. At the same time, public awareness increases and the company may cross a perception threshold. However, this increasing popularity is also observed by competitors who will also enter the market if the market prospects are attractive. During this second phase prices are frequently set such that the market share is maximized.
3. *Saturation Phase*. In the third phase, costs tend to decrease in general due to an optimized costs structure and an increasing market awareness. However,

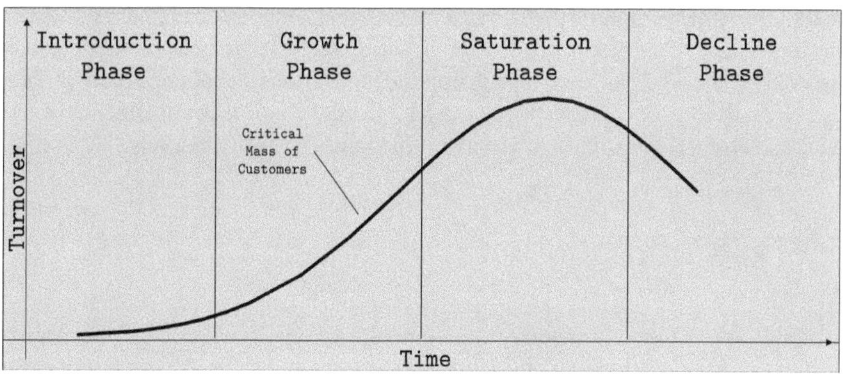

Fig. 3.1 Product Life Cycle.
Source: (Day 1981)

[4] Please confer also Sect. 7.2.1 for the different roles of adopters in customer networks during this product life cycle.

peak sales induce further competitors to cross the market entry barriers and to compete in the attractive market. Prices tend to decrease due to increasing competition. Strategies such as brand differentiation and feature diversification become important in the competition.

4. *Decline Phase*. The final phase is characterized by decreasing sales volume and declining prices. Consequently, the overall profitability decreases and profits depend decisively on the distribution efficiency in the customer network. The company has to either withdraw, update or replace the software in order to enter the product life cycle, anew (Box 1983).

It is important to note that the total cycle period in software markets is very short in comparison to conventional industrial products, such as cars (Dhalla and Yuspeh 1976; Grantham 1997).

3.3.2 Hypercompetition and Disruptive Innovations

Hypercompetition is a key feature of software markets that is closely related to the previously outlined short life cycle of software. *Hypercompetition* describes an intensive state of competition with a very high rate of change such that only the most adaptive firms survive. In software markets the competition is more intense than in other markets as the new type of customer requires quick, cheap, and highly customized services (D'Aveni 1994). Drivers of global software markets are quick customer changes, rapid technological changes and decreasing geographic boundaries. Hence, competitive advantages in software are not sustainable over long periods, but rather continuously created, eroded, destroyed and recreated. Consequently, it is possible to observe disruptive innovations in software markets.[5] There are two types of disruptions, which can be classified into low-end and new-market innovations. New-market disruptive innovations serve customers with previously unserved needs, whereas lower-end disruptive innovations aim at previously ignored customers. Both types of disruptions can lead to vital changes in customer networks and are, therefore, relevant to respective valuation in software markets.

3.3.3 Intangible Assets

Intangible assets are non-monetary claims on future benefits which cannot be seen or touched, as they do not have a physical embodiment (Müller 1990). In essence, there

[5] A *disruptive technology* is a technological progress, product, or service that overturns the existing dominant technologies (Bower and Christensen 1995). Please confer the close relationship to the tipping character explained in Sect. 3.3. Similarly, *disruptive innovations* are defined as new products in new markets (Bower and Christensen 1995).

exist two primary forms of intangibles, namely legal intangibles and competitive intangibles. Legal intangibles such as trade secrets, copyrights, patents and legal property rights are defensible in court. Competitive intangibles such as knowledge, human capital, or synergies are not legally ownable, but have a direct impact on the efficiency and productivity of a company. Both types of intangible assets are highly relevant for valuation in software markets as they can account for a large proportion of the total value (Lev 2001).

3.3.4 *Information Goods*

The intangible information goods traded in software markets have different characteristics than conventional industrial goods (Lev 2001). While tangible industrial goods typically have a very fast product life cycle, this is not necessarily true for intangible assets in software markets, which can even be regenerative.[6] In turn, the value of intangible assets in software markets can vanish, even instantaneously, if they become public property or outdated, e.g. software code. In such cases, the value depends on the product life cycle and the relative rate of innovations in customer networks. Both can vary significantly among various types of software products (Grossman and Helpman 1994).

3.3.5 *Nonrivalry*

Most tangible goods are rival assets, i.e. the scarcity of resources implies that opportunity costs resulting from consumption have to be considered. In contrast, software markets are primarily characterized by nonrivalry. Such nonrivalrous goods can be used by an infinite number of people in an infinite number of ways without harming the utility of any other person (Lev 2001). In other words, software can be classified as a nonrivalrous good due to the marginal production costs for duplication.

3.3.6 *Property Right Protection*

Property right protection is a vitally discussed phenomenon of global software markets. Despite property protection initiatives, it is rather difficult to prevent

[6] An illustrative example for regeneration is knowledge which increases if it is shared. Likewise, the value of data can increase over time, such as databases which become more valuable if the amount of data increases.

software piracy as large parts of the value creation process can be duplicated with marginal efforts (Lev 2001). Software pirates breach patents and trademarks by using software for personal or business purposes without the appropriate license. This behavior has a large impact on the business models of software companies. Despite increased copyright protection mechanisms, the Business Software Alliance announced for 2006 that the global total loss of software piracy grew by USD 15 bn from USD 25 bn in 2005 to USD 40 bn (BSA 2007).[7] But property right protection is not limited to software piracy in software markets. Another vital class of problems is related to the tacit knowledge of employees. In software markets social capital is very scarce and extremely difficult to protect as companies frequently cannot reinforce the copyright on their implicit knowledge if their employees switch to competitors or start their own businesses. In addition, depending on prices, service levels and switching costs it can also be difficult to protect the customer base in software markets (Shapiro and Varian 1998). All factors taken together illustrate that property right protection has a decisive influence on valuation in software markets (Prahalad and Hamel 1990).

3.3.7 Peculiar Cost Structure

Software markets have a peculiar cost structure that results from the outlined short customer life cycle.[8] Many software projects require substantial initial investments for the design and creation of software products in the market introduction phase, while replication and distribution costs are comparatively low.[9] Such large sunk costs imply a high degree of asset specificity which can be measured as the difference in asset value between the original purpose and that of an alternative use (Klein et al. 1978). The higher the asset specificity, the higher the risk profile of an industry (Lev 2001). In software markets the asset specificity and the corresponding risk profile is relatively high (Scherer et al. 2000). Variable costs in software markets are in turn low as the costs for reproduction and distribution of digital products are marginal, but incur the outlined copyright protection problems.[10] Hence, the role of developments in customer networks are also highly relevant from a cost perspective, as the number of customers has a decisive impact on the cost structure.

[7] It is important to note that the published statistics are subject to discussion as the Business Software Alliance has a biased perspective on software piracy. Founded in 1988, the Business Software Alliance is a group of large software makers that wants to stop copyright infringements.

[8] Please confer Sect. 3.3.1.

[9] The initial costs are also known as *setup* or *sunk costs*. Both terms describe costs which are not be recovered.

[10] Please confer Sect. 3.3.6.

3.3.8 Supply-Side Economies of Scale

In economic terms, the combination of high initial fixed costs and low marginal costs is also termed supply-side economies of scale (Liebowitz and Margolis 1999). In other words, the profitability of the software increases with rising volume due to the instant scalability if there are no capacity constraints. The setup and extension of the customer base is also effected by supply-side economics of scale as trust in a product is reinforced by a large customer base and a strong brand name. Growing market shares increase the popularity of the product and, thereby, reinforce the sale of further products, whereas average production costs decrease (Shapiro and Varian 1998). This mechanism underlines the vital importance of word-of-mouth referrals in software markets (Goldstein 1998).

3.3.9 Information Overload and Trust

The Internet is a source of information, but it also reinforces the risk of an information overflow. Hence, it underlines the importance of datamining tools and data navigation tools (Hecker 1998; Evans and Wurster 2000). Due to short product life cycles and the outlined hypercompetition, information flows in software markets are far more difficult to control than in other industries. At the same time, word-of-mouth referrals and trust become even more important in such a market environment (Goldstein 1998; Korb 2000). Another prerequisite of purchases in software markets is that customers trust the Internet as a distribution channel. Sensible information such as credit card information and personal data are exchanged during a purchase. In this dilemma, the signaling effect of brands can facilitate eCommerce, but the creation of trust requires time and resources. The interdependency between popularity and trust is another relationship that underlines the importance of word-of-mouth referrals in customer networks of software markets (Goldstein 1998).

3.3.10 Startup Companies

Startup companies are characteristic of software markets. As they frequently incur losses in the initial market phase, the terminal value represents frequently a significant part of the total value. At the same time, the historic data is frequently unavailable or very unreliable which increases the uncertainty of investments in such companies (Porter 2001, p. 64). Hence, venture financing of a software company is a investment in an highly uncertain environment in which terminal values have significant importance (Sturm 2003).

3.3.11 Network Effects

Network effects are demand-side economies of scale. They are defined as the change in benefits that agents gain from a good when the number of other customers changes who are consuming the same kind of good (Leibenstein 1950; Rohlfs 1974; Dybvig and Spatt 1983; Farrell and Saloner 1986; Katz and Shapiro 1992).[11] In other words, the size of the customer network is a factor that determines its value to other adopters as the membership of an individual increases the benefits of other network members. Network effects become particularly significant if a specific percentage of the overall population has subscribed.[12] Hence, it is a key business concern in software markets, to attract users prior to reaching critical mass, as even small changes in the customer network can have a vast impact on its value if they cross the critical mass (Economides 1996; Shy 2000a; Lev 2001). With an increase in size, the network becomes capable of attracting a wider user base. Such reinforcing feedback underlines the vital importance of word-of-mouth referrals (Goldstein 1998). Hence, it may be reasonable for a software company to provide early adopters with extrinsic motivations such as payments, fee waivers or referral boni. Such measures can be interpreted as investments in the customer network. But networks can also, in turn, become congested or saturated. In this case, additional users decrease the value of the network.

3.3.12 Reconsideration of Characteristics of Software Markets

All in all, the outlined aspects illustrate that network effects influence the price of software products, the pricing strategy of software companies and their respective values. Due to their vital importance, the role of network effects in valuations in software markets is a discussion to be expanded in the coming chapters. In summary, the profiling of modern software markets revealed their vital characteristics. They are relatively young dynamic markets for intangible, nonrivalrous information goods, services, and applications with short product life cycles. Supply and demand of software markets are vitally determined by their peculiar cost structure and the

[11] Please note the difference between network effects and *network externalities*, which are present if the market participants fail to internalize these effects. Even if individual users are not likely to directly internalize the effect of their membership, other agents such as the owner of a network may very well internalize them. In such a case, the respective network effects are no longer externalities. Unfortunately, the terminology has been some what mixed in literature (Liebowitz and Margolis 1994). Similarly, network effects are frequently mistaken for economies of scale, which result from business volume rather than interoperability. Therefore, demand-side and supply-side economies of scale are explicitly labeled if a distinction is required.

[12] Please note that this specific percentage is frequently termed the *critical mass* of a network (Economides and Himmelberg 1995). It is defined as the minimal non-zero equilibrium market coverage of a network good or service.

importance of trust. As a consequence of the outlined attributes software markets are complex in comparison to other markets. Further vital characteristic relevant to valuation in software markets is the low liquidity of assets and the vital importance of mouth-to-mouth marketing. At the same time, the variety of relevant characteristics underlines the importance of network effects in the business development of software companies. After having depicted valuation tools and software markets individually, both aspects are reconsidered from an integrated perspective in the subsequent chapter.

Chapter 4
Reconsideration of Valuation in Software Markets

> "Flexibility has value. While this statement is obvious at the conceptual level, it is surprisingly subtle at the applied level. Professional managers have long intuited that both operating flexibility and strategic flexibility [...] are important elements in valuation and planning decisions. But precisely how valuable is flexibility, and how can its value be quantified?"
>
> Mason in (Trigeorgis 1996)

The previous analysis outlined the challenge of precise valuation in dynamic software markets. In this chapter, various valuation approaches are compared with the requirements for valuation in software markets. According to this comparison, the real option approach is perfectly suited as it best accounts for the uncertainties and flexibilities in software markets. Since the real option approach can be implemented by various option pricing models, possible alternatives are matched with the requirements for valuation in software markets. As numerical simulations are identified as the preferential method, the modeling of the respective input parameters is investigated. Volatility and price of the underlying are identified as particularly important and challenging parameters. Hence, they are analyzed in the following part.

4.1 Reconsideration of Traditional Investment Valuation

The analysis of valuation in software markets raises the question which approach should be preferred. In order determine the optimal choice, all relevant valuation approaches are compared to the derived profile of software markets in order to account for the identified network effects.[1]

4.1.1 Reconsideration of the Asset Value Approach

The Asset Value Approach provides only limited contributions to valuation in software markets. Its retrospective and static nature does not take the flexibilities nor

[1] Please confer Sect. 3.3.

A. Kemper, *Valuation of Network Effects in Software Markets*,
Contributions to Management Science, DOI 10.1007/978-3-7908-2367-7_4,

the intangible assets into consideration, which are significant characteristics of software markets. Empirical investigations reveal that the value of respective flexibility options can represent a significant part of the total value (Keiber et al. 2002).[2] However, the Asset Value Approach is a very vague approximation for the liquidation value of a software company.

4.1.2 Reconsideration of the Market Value Approach

This approach is based on comparisons of similar market-formed prices. Despite of their popularity, a decisive limitation in the application to valuation in software markets is the required similarity of the compared assets or companies. Depending on the segment, software markets are liquid and provide sufficient data. However, it is frequently not possible to identify a twin asset with a similar risk profile (Trigeorgis 1996). In addition, market values are subject to a variety of factors identified by the research on behavioural finance, e.g. market inefficiencies, framing effects and heuristics, which are depicted in detail in the following Sect. 4.1.3 (Kahneman and Tversky 1979; Camerer et al. 2005).[3]

4.1.3 Reconsideration of the Discounted Cash Flow Approaches

In comparison to the market value approach, the DCF approaches account for more specific investment information (Copeland et al. 1996). Nevertheless, empirical and theoretical research reveals that they are subject to a variety of limitations.[4] General criticism stems from the underlying neoclassic assumptions, as research on behavioral finance challenge several underlying assumptions concerning information efficiency and rational behavior. Examples are market inefficiencies, framing effects and heuristics (Kahneman and Tversky 1979; Röder and Müller 2001).[5] The most relevant aspects for valuations in software markets are summarized in Table 4.1 (Shiller 2000; Müller 2003).

---

[2] Please confer Sect. 3.3.

[3] Please confer to the additional criticism depicted in Sect. 4.1.3.

[4] Please confer (Trigeorgis 1996) for an overview.

[5] The neoclassical model is criticized for assuming rational behaviour and information efficient markets (Fama 1970; DeBondt 1995). In this stylized world, it is impossible to gain an information advantage in information efficient markets based on an analysis of publicly available information (Möller 1985). While initial tests of the efficient market hypothesis seemed to confirm a weak or medium form efficiency, other investigations of pricing anomalies in capital markets contradict the previous findings (Fama 1970). Please confer (Copeland and Weston 1988) for further details on the neoclassic assumptions and (Camerer et al. 2005) for an overview on behavioral finance.

Table 4.1 Criticism of the neoclassical model

Researcher	Criticism	Content of Criticism
Hoppe (1930) Rosenfield (1976)	Attribution Bias	Success is attributed to capabilities while failure has external reasons
Kahneman and Tversky (1979)	Loss Aversion	Indifferent decisions are not considered to be equal
Tversky and Kahneman (1981)	Mental Accounting	Various mental accounts exist in parallel
Lichtenstein, Fischhoff and Phllips (1982)	Overconfidence Bias	Own prognosis is too optimistic
Fischhoff (1983)	Presentation Effect	Framing effect: Assessment depends on context
Arkes and Blumer (1985)	Sunk Cost Effect	Sunk costs have impact on current assessments
Frey (1986)	Selective Perception	Decision conform information is more likely to be considered
Unser (1999)	Risk Assessment Bias	Personal references influence decisions
Nofsinger (1999)	Herding Behavior	Group dynamic decision making
Shiller (2000)	Irrational Exuberance	Stock Market Bubbles

Source: Author adopted from (Müller 2003)

In addition to this general criticism, theory and practice have identified the following three limitations that are particularly relevant in software market valuations.

1. *Cost of Capital.* The cost of capital is a decisive variable in DCF approaches that is restrained by several constraints. A central restriction is the derivation of the risk-adjusted discount factor based on the CAPM or APM (Roll 1977). As the CAPM is a one-periodic model, rigid modifications are required if the capital structure or the risk profile of the software company change significantly over time. Another vital criticism of the CAPM concerns the market portfolio, and risk-free rates, which are theoretical constructs that have to be approximated in practice (Roll 1977; Brach 2003).
2. *Negative Cash Flows.* Software companies require considerable initial investments, pay low or even no dividend payments, have a short financial history and are subject to exponential growth. Even if they have high growth potential, they

frequently also have negative cash flows. Therefore, firms may not be considered solvent, despite substantial but illiquid intangible assets (Schäfer and Schässburger 2001). In such situations, the choice of the risk-adjusted discount factor and the expected cash flows decisively influence the outcome of valuation in software markets.

3. *Managerial Flexibilities*. Conventional valuation approaches assume fixed investment strategies based on a given set of information. Hence, the quality of conventional valuations is limited by the initial predictions. In particular, subjective expectations and assumed risk-adjusted discount rates are parameters that are very sensitive to errors (Born 1995; Ballwieser 1995). Moreover, this perspective ignores the value of managerial flexibility, which may emanate from new information during the investment horizon. Sometimes management gains the flexibility to adapt the strategy to the changing environment as uncertainty gradually resolves over time at the expense of an option premium. Such an asymmetric payoff structure implies that the upside potential of the investment increases, while the downside is limited to the initial investment costs. If, in turn, the flexibilities are ignored, the investigated assets or companies are chronically undervalued (Bellinger and Vahl 1984; Trigeorgis 1996; Brach 2003). Consequently, boundary projects are not conducted, as banks and financial investors do not provide an adequate level of capital.

All in all, conventional valuation approaches are restricted by a variety of limitations, most of which are related to the adequate consideration of risk. Consequently, they can lead to inefficient investment decisions. The real options approach seems to be suitable for valuations in software markets as it allows one to overcome some of the shortcomings by identifying, disaggregating, and valuing relevant flexibilities. As these factors are particularly important in software markets, it is reasonable to consider the application of a real options approach. This choice is also supported by other more general valuation guidelines as software markets are highly uncertain and provide high managerial flexibilities (Hommel 1999). The application of real options to valuations in software markets is also supported by related successful applications which reveal the opportunities but also the limitations of this approach. The respective research is reviewed in the following section (Fig. 4.1).

4.2 Reconsideration of Real Options Approaches for Valuation in Software Markets

A review of the literature reveals that a number of researcher have written on the contributions of real options to various aspects related to valuation in software markets. While most papers have a focus on general IT investments, there exist also some studies that are more specifically related to valuation of companies operating in software markets:

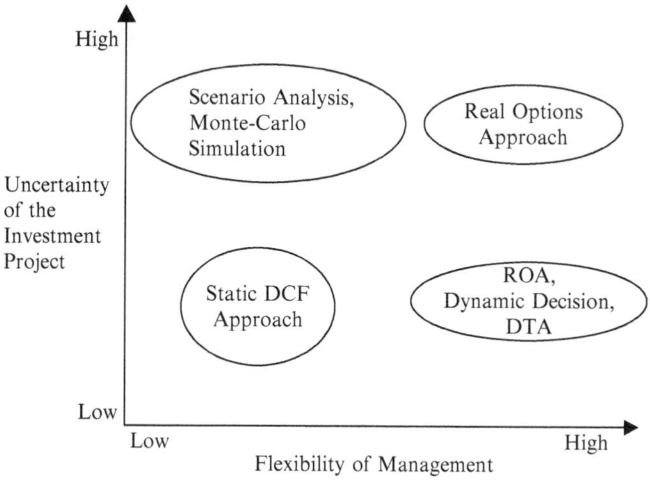

Fig. 4.1 Valuation approach guideline
Source: Hommel (1999)

Innovative Companies. In the real option literature there exist several examples that demonstrate the possibility of real option valuation for companies under comparable business conditions, e.g. Internet companies, biotech companies, startup companies (Willner 1995; Schwartz and Moon 2000; Kellogg and Charnes 2000). The employed valuation approaches vary with respect to the level of detail, to the applied option pricing model and with respect to the approaches used to derive the option parameters development of the underlying and volatility of the option.

Strategic Value of IT Investments. Investigations on the application of option models in IT investment decision making roots back to a seminal paper of (Dos-Santos 1991). In this paper a Margrabe exchange option model is applied for valuing a new IS project in order to account for the learning and experience derived by the project for further projects (Margrabe 1978). This paper incited a stream of investigations on the option-like characteristics of information technology investment projects (Clemons 1991; Kambil et al. 1993; Kumar 1996; Chalasani et al. 1997). Based on such insights further research apply the real options approach in order to develop a risk management strategy for technology investment risk (Benaroch and Kauffman 1999; Benaroch and Kauffman 2000; Benaroch 2002).

Modularity and Platform IT Investments. Flexible responses to the increasing pressure for information systems to be readily adaptable to changing business processes contain value. Hence, modularity and extendability is incorporated into quantitative IT investment decision models with real options (Baldwin and Clarke 2000). Accordingly, software design concepts were user to develop a theory of modularity in design based on real options.

Software Economics. In another stream of research the focus is more specifically on the flexibility of software by a platform design approach. This implies the modularity in design, software economics, and software dependability with the goal to optimize the economic outcomes of software development (Sullivan et al. 1999).

Software Platform Investments. Another stream of research investigates platforms investments of companies into software (Taudes 1998; Taudes et al. 2000). Accordingly, the modularity and extendability of software is incorporated into quantitative IT investment decision models with real options.

Technology Competition and Investment Timing. Another stream of research applies real options in order to determine the optimal timing for an IT investment (Kauffman and Xiaotong 2005). Accordingly, there are exist closing windows of opportunity for IT investments. Accordingly, real option approaches are applied in order to determine the optimal time to invest into a new technology which is influenced by stranding costs on the one hand and increasing opportunity costs on the other hand.

In summary, there are several examples that demonstrate the suitability of real option valuation for companies under comparable business conditions, e.g. Internet companies, biotech companies, startup companies and even for IT and software related issues. The ability of the real option approach to account for the uncertainty of future market developments expressed in a highly volatile business is a recommendation for the valuation in software markets. The impact of managerial flexibility and the risk inherent in software companies provide significant advantages of the real options approach over traditional analysis methods. In those traditional approaches, cash flows are perceived in rigid scenarios and the risk adjustment does not account for the investment situation in which the CAPM does not permit to derive properly the risk associated with the software markets. The real options approach, in contrast, captures the value of growth and flexibility resulting from customer networks in addition to the net present value of the cash flows yielding the extended company value. But despite of considerable research efforts there are no research contributions that investigate the valuation of companies operating in software markets based on a customer network-centric perspective. Moreover, it is outlined that the choice of an option pricing model is of primary importance. As various option pricing models are available to implement the real options approach, the most relevant models for valuation in software markets are compared in the following section.

4.3 Reconsideration of Option Pricing Models for Valuation in Software Markets

The purpose of this section is to identify the most suitable option pricing model for valuation in software markets with the real options approach. There are two types of option pricing models with various subclasses. While analytical approaches are based on closed-form solutions, numerical solutions are derived by numerical

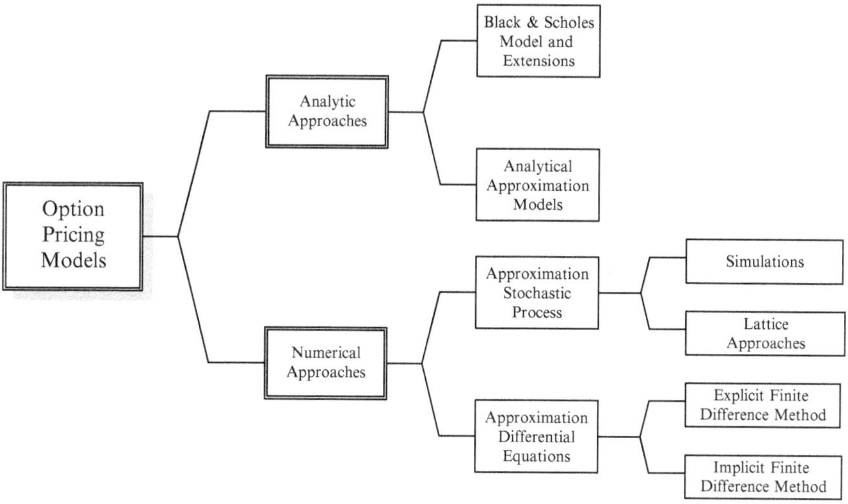

Fig. 4.2 Typology of option pricing models
Source: (Hommel 1999)

approximations.[6] Analytical approaches are subdivided into the classical Black and Scholes model with extensions and analytical approximation models. Numerical approaches, in turn, comprise approximations of stochastic processes with simulations or lattice approaches and approximations of differential equations based on implicit or explicit finite difference methods. The models can be classified as depicted in Fig. 4.2.

In order to determine the most suitable model for valuation in software markets, all approaches are compared based on relevant criteria (Geske and Shastri 1985). A research of the respective research indicates that the following criteria are most relevant to valuation in software markets:

1. *Transparency*. Valuation in software markets has to be transparent. The degree of the required and desired transparency of the option pricing approach has to account for the target group of the valuation. In general, intransparent valuations are more likely to be rejected.
2. *Complexity*. Valuation in software markets has to be simple. An increasing degree of complexity, e.g. caused by compound options or interacting options, induces an over-proportional increase in computational complexity. Thereby, models become increasingly inflexible for extensions.
3. *Precision*. Valuation in software markets has to be precise. The precision of most approaches is inversely related to the outlined complexity criteria. While some option pricing approaches can be interpreted as rules of thumb that deliver a

[6] Please confer (Hommel et al. 2003; Baecker and Hommel 2004; Schulmerich 2005) for further information.

first idea of a value, other methods are more accurate, but frequently also more complex. Therefore, it is an important goal in the selection process to consider the tradeoff between complexity and precision.

4. *Prerequisites*. Valuation in software markets should not be based on restrictive assumptions and have to account for the available data. All option pricing models are based on implicit assumptions. Analytic methods, for example, require that it is possible to construct a portfolio that is perfectly correlated with the underlying. Otherwise, tracking errors have to be reconsidered as this would cause an overestimation of the option value otherwise (Amram and Kulatilaka 2000, p. 10).

Based on the outlined criteria, the approaches can be compared with respect to their capabilities and limitations for valuation in software markets.

4.3.1 Analytical Solution Methods

The Black-Scholes formula is a very popular set of closed-form partial differential equations that is used to value options. Accordingly, the value of the option is formulated as a function of time and of an underlying based on risk-free valuation combined with the Law of one Price (Black and Scholes 1973). The precision of analytical solutions increases with the number of intervals (Hommel 1999). In comparison to other approaches, analytical approaches have the advantage of a higher calculation efficiency due to the ease of application. At the same time, it is necessary that a solution exists. However, it is frequently difficult or even impossible to derive an analytical solution. This is particularly relevant for valuation in software markets, due to the identified interdependencies of customer networks (Merton 1995). Therefore, alternative analytical approximations are investigated.

4.3.2 Analytical Approximation Methods

A variety of approaches are allow to approximate analytically solutions for the option. For example compound options are approximated by analytic polynomial approximations (Geske and Johnson 1984). Other researchers suggest quadratic approximations in order to value options (Barone-Adesi and Whaley 1987). A third group of researchers investigated other heuristics for problem-specific analytic approximations (Blomeyer 1986). With respect to valuations in software markets, the analytical approximations are restricted by similar boundaries as purely analytical solutions. But in essence, they are not suitable for capturing complex interactions in the underlying customer network. Hence, numerical approaches are explored in the following.

4.3.3 Lattice Approaches

The binomial tree model is a binary numerical lattice approach that interprets the development of the stock price as a time and state discrete multiplicative binomial process.[7] The standard valuation procedure is recursive and requires one to identify the value of the option over any time interval with two trees. While the first tree describes the development of the underlying, the second tree models the value of the option with respect to the development expressed in the first tree. In this approach, time is deconstructed into a finite set of intervals in which each state leads to either an upward or a downward development. A series of such time steps is termed a *binomial process*. The respective option value is derived by a replication portfolio, i.e. a mix of securities and risk-free assets. As both have identical cash flows, they are supposed to have a common price.[8] Lattice approaches are in general very intuitive and more flexible than most other approaches, as far as issues such as handling of stochastic processes, option payoffs, early exercise decisions and several underlying variables, etc. are concerned. Another main advantage in comparison to traditional valuation techniques is that the value is independent from a personal risk preference. Hence, real-world probabilities are not required. Valuations of multiple period options are possible based on the roll-back method. As the terminal values are known, the values in previous periods are derived by discounting the terminal value at the risk-free rate. In comparison to other option pricing models, another advantages is that American options with early exercise opportunities can be valued, as the values of an early exercise are tested in each period. In turn, lattice approaches are less accurate than alternative approaches and the computational effort increases exponentially with the number of considered intervals. But lattice approaches are not capable of handling more than one starting price at a time. In such cases the log-transformed binomial lattice method can be applied in order to resolve emerging problems of consistency, stability, and efficiency as it allows one to model options with multiple option investments, exercise prices, and interactions (Trigeorgis 1991). For this purpose the Brownian motion is approximated by a discrete process with the same mean and variance at each discrete state, which is defined in logarithmic transformed state and time units. Consequently, the stability and the consistency of the approach increases. It delivers a better estimate in fewer steps which reduces the costs of computation. Dividends representing exogenous factors, such as competitors, can be implemented in contrast to analytical closed-form solutions. Nevertheless, the approach is less intuitive as the logarithmic transformation can be seen as a black box. All in all, the decision tree approach is intuitive and is a flexible tool in order to determine the optimal exercising strategy,

[7] While the idea roots back to Sharpe, the first algorithmic implementation dates back to (Cox et al. 1979) and (R. J. Rendleman and Bartter 1979).

[8] This idea is also termed the Law of One Price. Accordingly, there exist no arbitrage opportunities in efficient and perfect capital markets (Merton 1976).

as competition and other factors such as compound options can be incorporated into the model (Bockemuehl 2001; Lucke 2001).[9]

4.3.4 Finite Difference Methods

Finite Difference approaches are time and state discrete approximations of the continuous differential equation for the option value, as the partial differentials are substituted by finite differences (Geske and Shastri 1985). The resulting set of difference equations is solved by a backward induction approach (Brennan and Schwartz 1978). In essence, the previously outlined binomial tree approach can be interpreted as a special case of the finite difference approach, in which the explanatory power of the binomial approach is higher, while finite difference methods are more accurate (Geske and Shastri 1985). Finite-difference methods are capable of valuing European as well as American options. These methods are the most efficient, if a set of start values of options have to be calculated. But they can cope also with multiple state-variables in a multidimensional grid, although such solutions are not very efficient.

4.3.5 Numerical Integration

Numerical integration incorporates the same idea as the finite difference methods. If an integral cannot be solved analytically, it is depicted as a sum of its parts (Baecker and Hommel 2004). Accordingly, the value of an option is interpreted as an integral over the stock price at maturity with a respective density function, which is replaced by discrete state probabilities of the underlying (Jarrow and Rudd 1983). A comparison to other approaches reveals that it is dominated by the binomial approach as it has no significant advantages, except for a conceptual problem with the transmission of the density function into the discrete probabilities (Parkinson 1977).

4.3.6 Simulations

Simulations can be applied in order to value options by generating the relevant risk-neutral probability distributions of the expected cash flows. Most common are *Monte-Carlo simulations* that date back to the research of Hertz on valuation under uncertainty and its application to option pricing by Boyle (Hertz 1964; Boyle 1977). In essence, the core algorithm is the generation of random numbers in order to

[9] Further details on the Binomial tree approach are depicted in (Hull 1989).

determine the distribution of a stochastic model. It is applied in option pricing to determine the volatility of an option. Monte-Carlo Simulations, can model even complex stochastic processes at low costs, but their contribution to the valuation of American options is limited, due to the early exercise possibilities of the owner (Hull 1989). Simulations allow one to value options in a simple and flexible manner, even in the face of complex path-dependent payoff structure of the underlying. Even multiple state-variables can be handled. Unfortunately, the forward-moving approach is not applicable to American options with an early exercise opportunity. Therefore, it is difficult to determine an optimal exercise policy as it is not possible to integrate intermediate decisions. In this case, finite-difference methods and lattice approaches based on a backward programming process are preferable.

4.3.7 Comparison of Option Pricing Models

The following table contains a comparison of the relevant option pricing models with respect to valuation in software markets along the most relevant decision criteria transparency, precision, complexity and knowledge prerequisites.[10] Their suitability for valuations in software markets is determined by matching the elaborated prerequisites their respective profiles. The respective scores range from 0 meaning rather negative to the best score + + + (Table 4.2).

The comparisons of the option valuation approaches reflects that lattice approaches are very transparent and versatile, but they are also less precise and more complex than other approaches. Finite differences and numerical integration are similar. They are more precise and more efficient than most other techniques, but they are also intransparent, complex and have a lot of prerequisites. Analytical solutions are precise but also less transparent than other approaches and have considerable requirements in order to be solvable. Their approximations are less precise, but they are also less complex than the analytical solutions. Simulations are identified to be the most suitable candidates with respect to valuation in software markets. They are sufficiently transparent, very precise and scalable with respect to the required level of detail and available data. This decision to choose simulation for an implementation of real option valuations in software markets is supported by the guidelines developed by (Amoco 2000). Accordingly, simulations are perfectly suitable tools for valuations in highly uncertain environments if the market liquidity is restricted, as is the case in software markets. All in all, the outlined evaluations reveal that numerical simulations are considered the preferential option pricing approach for the real option valuation in software markets. For this reason, their implementation is analyzed in the following part of the book (Fig. 4.3).

[10] Please compare (Geske and Shastri 1985) for a derivation of the performance criteria of option pricing models.

Table 4.2 Comparison of option pricing models

Valuation Approach	Transparency	Precision	Complexity	Prerequisites	Overall Suitability
Analytical Solutions	0	+++	++	+	++
Analytic Approximations	0	++	+	++	+
Lattice Approaches	+++	++	++	+++	++
Finite-Difference	0	+++	0	0	+
Numerical Integration	0	+++	0	0	+
Simulations (e.g. Monte Carlo)	++	+++	+++	++	+++

Source: Author

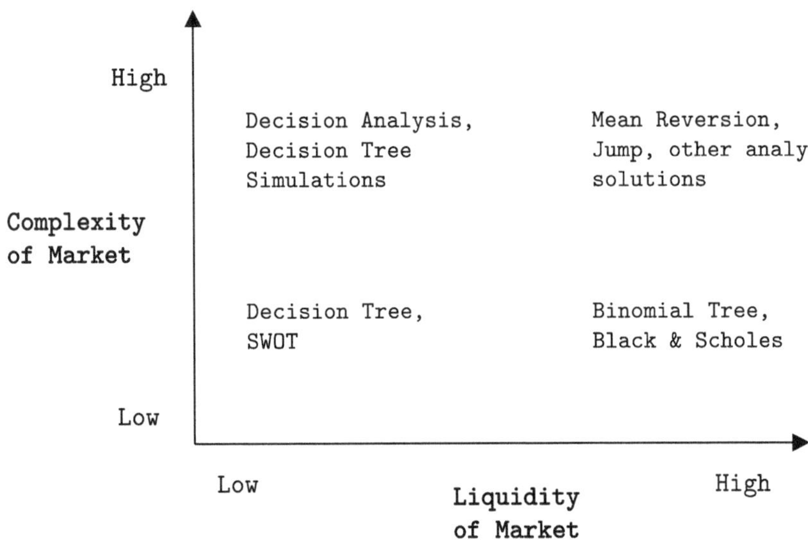

Fig. 4.3 Option pricing tool guideline
Source: (Amoco 2000, p. 15)

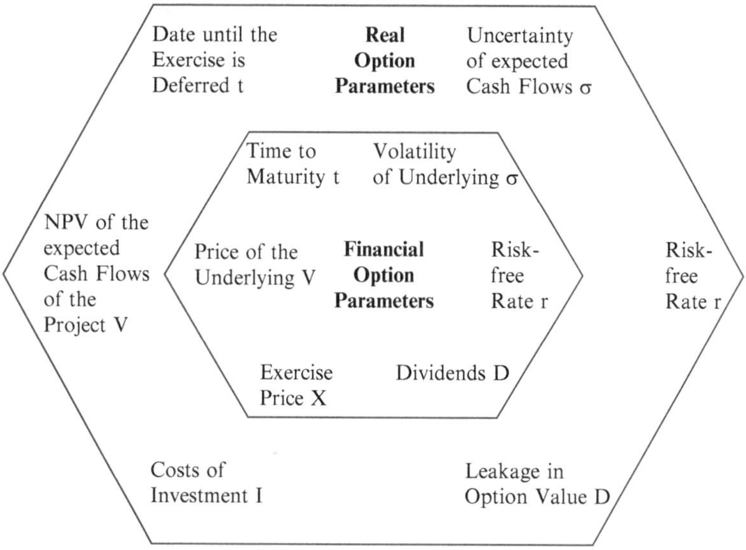

Fig. 4.4 Financial and real option parameters
Source: Trigeorgis (1996)

4.4 Parametrization of Real Option Models

After the selection of the real options model, the next step in the valuation process is its parametrization.[11] Similarly to financial options, the value of Real Options is determined by six parameters (Trigeorgis 1996). While some of the outlined parameters are frequently observed, others are more difficult to obtain (Hommel and Pritsch 1999; Schulmerich 2005) (Fig. 4.4).

1. *Price of the Underlying* V. The payoff is the present value of the expected cash flows. If traded, the value of the underlying can be determined instantaneously, otherwise it is approximated by discounting the expected cash flows at a risk-adjusted discount factor. Increasing present values have a positive impact on the value of an option.
2. *Exercise Price* X. The exercise price corresponds to the present value of the initial investment costs, if the option is exercised. This parameter reduces the value of the option and can be more easily determined than other parameters.
3. *Time to Maturity* t. This parameter determines when the flexibility expires. In general, a longer time to maturity increases the value of the option. As real options are frequently not contractually fixed, their beginning and end have to be approximated. The time to maturity can be easily determined if the option is driven by exogenous restrictions, such as patents. If competition is involved,

[11] Please confer Fig. 8.2 in Sect. 8.4.

it is more difficult to determine the time to maturity due to game theoretical interdependencies, which are usually modeled by adjusting the dividend of the option.
4. *Volatility* σ. Volatility is a measure of the related uncertainty. Contrary to intuition, increasing uncertainty raises the value of an option. It is one of the most important and challenging parameters in the modeling of options for valuation purposes.
5. *Risk-free Rate* r. This parameter is a measure of the risk-free opportunity cost of capital. Increasing opportunity costs imply a decreasing option value. The discount rate for the risk-neutral valuation has to account for several requirements. A common time to maturity is required as well as a common currency. Since interest rates may vary over time, a long-term time horizon is required as well as interest-rate models are required.
6. *Dividends* D. Dividends are cash-in and cash-out flows representing leakages in the value of an option. They lower the value of an option and represent opportunity costs if the option is not exercised, e.g. due to competition.

As the specification of the volatility and of the underlying are or primary importance in the parametrization of real option models, they are analyzed in greater detail in the following sections.

4.5 Specification of the Volatility

A variety of approaches are available to determine the volatility of an option. But it is frequently impossible to observe the volatility directly, it is instead derived based on the following considerations (Fig. 4.5).

1. *Heuristic Volatility*. Research suggests that the heuristic volatility ranges typically between 30 percent and 60 percent for intra-company projects (Luehrmann 1998). Other investigations find a broader empirical range particularly for aggregated companies (McDonald and Siegel 1986, p. 717). It is important to note, however, that both heuristic estimations are rules of thumb which can be susceptible to distortions.
2. *Implied Volatility*. Market values represent the primary choice to determine the required volatility of options. Accordingly, the implicit volatility of market values can be obtained by an inversion of the pricing process if the asset is traded and has a high liquidity, i.e. for homogeneous goods and services or natural resources. The basic idea is that it is not difficult to determine the volatility assumed by the market if the corresponding market prices can be observed. If the asset is not directly traded, the respective volatility of a twin security can be applied if the respective data is available (Davis 1998, p. 723). In this case, adoptions may be necessary in case of imperfect tracking, but the values are reasonable approximations of the volatility if both assets have similar characteristics. The idea is that the market prices of the traded options contain implied

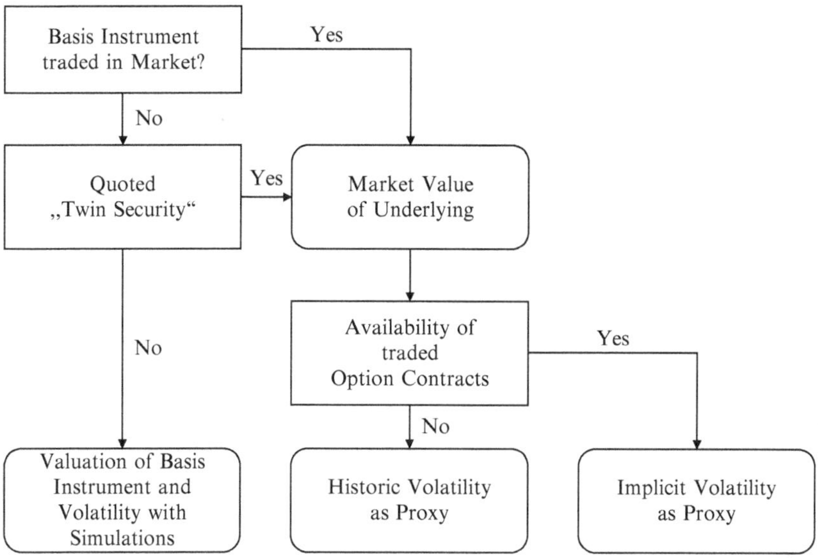

Fig. 4.5 Determination of volatility
Source: Hommel (1999)

estimations of the respective volatilities. Primary application of implied volatility are resource-based projects.

3. *Historic Volatility*. The analysis of historic time series is another way to determine the volatility of an option based on market values. It is derived from an extrapolation of historic data. Based on this idea, historic data is used to approximate the estimated volatility by $\hat{\sigma} = \frac{s}{\sqrt{\delta t}}$ where

$$s = \sqrt{\frac{1}{n-1} \cdot \sum_{i=1}^{n} (u_i - E[u_i])^2}, \qquad (4.1)$$

with $u_i = ln\left(\frac{S_i}{S_{i-1}}\right)$, with S representing the value of the underlying, with the standard deviation σ, with n equal to the size of the sample, e.g. daily data 180, and with u as the return.

4. *Simulation of Volatility*. In addition to the outlined approaches, a simulation of the expected cash flows can be applied in order to derive the standard deviation of the inherent risk. Such simulations are frequently performed based on a Monte-Carlo approach, according to which a synthesized probability distribution is derived by multiple iterations. Thereby, simulations provide numerical estimates depending on the parametrization of the model (Davis 1998, p. 730). All in all, simulations are efficient second-best solutions that are capable of accounting for complex features of diffusion processes, such as network effects, which yield far more reliable data if they are properly designed and specified.

4.6 Specification of the Underlying

In addition to the volatility, the development of the underlying is another challenge in the parametrization of real options. Accordingly, the development of the underlying is frequently described as stochastic processes in probability distributions which are mathematical descriptions for the behavior of random variables (Cox and Ross 1976; Merton 1976; Cox and Rubinstein 1985).[12] They should be applied if single expected values are inadequate and imply a considerable level of complexity, as most arithmetic and algebraic operations cannot be applied.

4.6.1 Probability Distributions

Depending on the nature of the underlying process, various probability distributions are distinguished. The most relevant probability distributions include the following three types.

1. *Basic Discrete Distribution.* A discrete distribution is a countable number of discrete outcomes. The binomial distribution is an important discrete probability distribution describing the frequency of events, in which only two states are possible.
2. *Continuous Distribution.* A continuous distribution is characterized by events over a continuous range. Standard distribution with normalized parameters $\mu = 0$ and $\sigma^2 = 1$ are frequently applied in continuous time finance, as differences in the price logarithms describe the development of stock price movements, as the natural logarithms of the relative stock prices are assumed to be log normally distributed. In continuous distributions the standard deviation and the mean range from negative to infinity. Continuous time and state random walks are frequently applied in option pricing (Jarrow and Rudd 1983).
3. *Lognormal Distribution.* A lognormal distribution is has a standard deviation and a mean which defined between zero and infinity.

A comparison reveals that a discrete random variable has a finite set of values, or an infinite countable set of values, in contrast to continuous random models. A substantial prerequisite is that the underlying is traded continuously, which is an acceptable condition in the valuation of financial options, but critical for real option valuation.[13]

---

[12] A stochastic process is a set of sequential states of a random variable X at a specific time t which is determined by a probability distribution p of the moments T. Consequently, the random events x_t are within an interval $[a_T, b_T]$. Stationary processes have constant characteristics within a certain time horizon, while they diverge in non-stationary models. Systematic trend developments are termed *drift*. Please confer (Dixit and Pindyck 1994) for further details.

[13] Please confer (Hull 1989) for a discussion of the problems related to the liquidity of the underlying.

Continuous time models can be approximated by specifying the parameters u and d in binomial models as these converge in the limit to the continuous version (Cox and Ross 1976).

4.6.2 Stochastic Processes for Valuation in Software Markets

In valuations the development of the underlying is frequently assumed to follow a Geometric Brownian motion, i.e. a Wiener process with a drift rate that represents the expected rate of return and with a variance expressed in the volatility of the underlying (Hull 1989, p. 223). However, there are also other stochastic concepts that are applied in option pricing theory. Therefore, the following section contains an overview on stochastic processes that are relevant to valuation in software markets (Hull 1989, p. 446).

4.6.2.1 Markov Process

A Markov Process is a stochastic process, in which the probability distribution of x_{t+1} depends solely on x_t and not on prior states. Transitions between states are described by transition probabilities of various stochastic processes. Markov-Processes can be subdivided in diffusion processes, jump processes and jump-diffusion processes which will be depicted in the following subsections.[14]

4.6.2.2 Wiener Process

A frequently applied process to model the stochastic development of a stock is the continuous time Wiener process which is derived from the underlying theory on Markov or Levy processes. The Markov process is a stochastic process in which the historic development of the process is irrelevant for the further development as the probability of future events are independent of the past behavior of the system. The Levy process, in contrast, is a continuous stochastic process with stationary independent increments. Basic types of the Levy process are the Wiener process and the Poisson process. The Wiener process is a stochastic diffusion process with the particularity of a mean change of 0 and a variance of 1 per year. A geometric Brownian motion is a generalized description of growth and diffusion processes, that are expressed by a drift and a variance rate which are time dependent and depend on the development of the underlying. The Ito process, in turn, is a generalized Wiener process with a drift and a variance rate that are functions of the underlying variable and of time. Accordingly, the expected value of the underlying

[14] Please confer 4.6.2 and (Hull 1989) for further details.

is supposed to depend on a contemporary value and that the changes over any finite time horizon are normally distributed, while, in turn, changes within a specific time interval are assumed to be independent of other time intervals (Merton 1995, p. 121). In analytical solutions the value of the underlying S is frequently described by a Wiener Process, i.e. a Geometric Brownian motion with drift. Formally, the Geometric Brownian Motion is defined as

$$\frac{dS}{S} = \alpha dt + \sigma dz. \tag{4.2}$$

Changes in value are expressed with the Itôs Lemma (Ito 1951)

$$dC = \frac{\partial C}{\partial t} dt + \frac{\partial C}{\partial S} dS + \frac{1}{2}(\sigma S)^2 \frac{(\sigma)^2 C}{\partial S^2} dt. \tag{4.3}$$

In this formula dS is the only stochastic variable. After the determination of the characteristics of the stock and the option, the price of the option is determined with a risk-free hedging portfolio. This allows one to derive the fundamental partial differential equation of contingent claims pricing in which the stochastic factor dz is eliminated

$$\frac{\partial C}{\partial t} = rC - rS\frac{\partial C}{\partial S} - \frac{1}{2}\sigma^2 S^2 \frac{\partial^2 C}{\partial^2 S^2}. \tag{4.4}$$

After transformations of the Black and Scholes formula, the value of the option can be derived by solving for C if the specific boundary conditions are considered

$$C(V, X) = SN(d_1) - Xe^{-rt} N(d_2), \tag{4.5}$$

with

$$d_1 = \frac{ln(\frac{S}{X})}{+} r + \frac{\sigma^2 2t}{\sigma\sqrt{t} = d_1 - \sigma\sqrt{t}} \tag{4.6}$$

and

$$d_2 = \frac{ln(\frac{S}{X} + (r - \frac{\sigma^2}{2})t)}{\sigma\sqrt{t} = d_1 - \sigma\sqrt{t}} = d_1 - \sigma\sqrt{t}. \tag{4.7}$$

The first term represents the value, if the option is in the money with $S_T > X$. The distribution function $N(d_1)$ equals the hedge-ratio n,i.e. the amount of stocks in the option portfolio. Term two represents the present value of the exercise price with $N(d_2)$ as the required financing. The analytic solution is a special cases of discrete models (Cox et al. 1979). While the multiplicative binomial process converges for $n \rightarrow \infty$ toward a log-normal probability distribution, it converges for many periods toward a standard distribution. As there frequently exist no analytical solutions, numerical approaches are applied. Examples are the lattice approach including the binomial tree model and the Monte-Carlo simulation in order to approximate the stochastic process of the underlying value (Hull 1989). Modifications of stochastic processes are applied if they describe the relevant development more accurately:

4.6.2.3 Mean-Reversion Process

Mean-reversion is a property of a stochastic process to remain near or to return over time to a long-run average value (Wilmott 2000). While interest rates and implied volatilities tend to exhibit mean reversion, exchange rates and stock prices frequently follow other patterns. But such properties may only reveal itself over very long horizons, which makes it difficult to spot. Therefore, the decision to model a quantity with a mean reverting stochastic process is frequently based on empirical observations in combination with theoretical reasoning.

4.6.2.4 Diffusion Process

Diffusion processes describe a continuous development of stochastic variables, where the developments are not assumed to be normally but log normally distributed, as values cannot be negative (Goldenberg 1991). The parameters u and d are adjusted to

$$u = e^{\sigma\sqrt{\frac{t}{n}}}, \tag{4.8}$$

and

$$d = e^{-\sigma\sqrt{\frac{t}{n}}} = \frac{1}{u}. \tag{4.9}$$

4.6.2.5 Jump Process

Jump processes are characterized by a discontinuous development of the analyzed random variable, where unpredictable moments such as new information or an external crisis induce jumps to another reference level (Merton 1976). Events have a unique impact, which influences the development at some point in time, but which has no further impact on future trajectories. In a binomial model, the parameter u and d are modified to $u = u$ and $d = e^{\zeta \frac{t}{n}}$ with ζ as the jump parameter determining the jump width of the underlying, the remaining time t over n intervals. For $n \to \infty$ results a logarithmic transformed Poisson distribution.

4.6.2.6 Jump-Diffusion Process

The jump-diffusion process combines the characteristics of a Jump Process with those of a Diffusion Process and is distorted by irregular lognormally distributed jumps (Cox and Rubinstein 1985; Hull 1989). Binomial models can be applied in order to model these processes. In such a model the parameters are specified as

$$u = e^{\hat{\sigma}\sqrt{\frac{t}{n}}} \tag{4.10}$$

and

$$d = e^{-\hat{\sigma}\sqrt{\frac{t}{n}}} = \frac{1}{u}, \tag{4.11}$$

with

$$\hat{\sigma} = \sqrt{\sigma^2 + k\vartheta^2}, \tag{4.12}$$

where $\hat{\sigma}$ is the total volatility of the underlying, ϑ is the jump component, σ is the volatility of the diffusion component, k are the jumps per year, t is the remaining time to maturity and n is the number of intervals per year (Sick 1989).

4.6.2.7 Reconsideration of Stochastic Processes for Valuation in Software Markets

Previous review reveals that the description of the development of the underlying is a central part of an option valuation approach. As a variety of stochastic processes are available for valuation in software markets, a crucial issue is to determine the most adequate approach. Within the outlined variety of process models it is important to note that the jump diffusion and the mean-reversion model are frequently applied for valuation in software markets.

4.7 Challenges of Valuations in Software Markets

After profiling the software markets and the overview of relevant investment valuation concepts it is possible to summarize the main challenges for valuations in software markets. As customer networks influence decisively the generated cash flows of the business model, their valuations represent a central challenge. Investigations indicate the central role of network effects in the customer base in valuation models. Hence, it is necessary to measure and quantify them. The outlined research underlines the importance of Real Option Valuation for valuations in software markets based on software market simulations, but several obstacles prevent its broad application. One of the most important implementation barriers is the derivation of the respective option valuation parameters. The profile of software markets revealed that most software companies are young and exhibit large growth potentials, but due to the short company history, there is frequently not much historic financial data available if any. Moreover, the high volatility of cash flows earned by software companies is a central theme in software market valuations. More precisely, the review of the specification requirements of a real options approach revealed that the volatility and the price of the underlying are the most challenging but nevertheless decisive option parameters. In order to overcome such implementation barriers, an innovative approach to modeling network effects for valuations in software markets is suggested in the subsequent part. Hence, numerical network simulations of the software market are designed in order to account for the identified characteristics of software markets and their diffusion behavior. Hence, network effects are investigated in the subsequent part on a finer level of detail in order to design a network effects framework for valuations in software markets.

Part III
Modeling Network Effects in Software Markets

As prior investigations revealed that network effects are determinants of customer networks, and since these effects are frequently ignored in conventional valuation approaches, the subsequent investigations intend to close the identified research gap. The goal is to design an integrated network effects framework for valuations in software markets by integrating network effects into corporate financial valuation. For this purpose, the relevant literature on network effects is reviewed in the first step, before customer network-centric valuation is introduced. This view considers the description of customer networks and the decomposition of cash flows based on software market models as crucial valuation issues in software markets. In a next step analytical and numerical adoption and diffusion models are investigated in order to describe developments of customer networks in software markets more precisely. All findings are integrated in a network effects framework for valuations in software markets. The goal of this framework is to use the additional network theoretical information of customer networks in order to enhance the quality of valuations in software markets. The part closes with a reconsideration of the developed framework.

Chapter 5
Network Economics in Software Markets

> "The industrial economy was populated with oligopolies [...] In contrast, the information economy is populated by temporary monopolies. Hardware and software firms vie for dominance, knowing that today's leading technology or architectures will, more likely than not, be toppled in short order by an upstart with superior technology."
>
> (Shapiro and Varian 1998)

The prior analysis revealed that network effects in customer networks are valuable drivers of growth opportunities in software markets. Therefore, the literature on network economics is reviewed in the following chapter. First some fundamental principles of network economics are depicted before a typology of network effects is suggested. Then, the most relevant characteristics and dynamics of software markets are outlined, before insights of empirical research are reviewed. Finally, network effects are investigated on a finer level of detail in selected software market segments. A reconsideration of network effects in software markets concludes this chapter.

5.1 Principles of Network Economics

Network effects are defined as the changes in decision variables of an economic agent, such as benefits, are based on choices of other agents consuming similar goods (Liebowitz and Margolis 1994). They occur in network effect industries such as telecommunication, transportation, electricity, banking, and health care (David 1985; Arthur 1989; David and Greenstein 1990; Economides and White 1993; Arthur 1996). Positive network effects in software markets are the additional utility that consumers gain, if the total number of software customers increases. This perspective allows one to separate the benefits of consumers into two distinct parts as network goods provide a direct utility and a derivative network utility that are a function of the size of the installed base (Katz and Shapiro 1985).[1] In other words, the size of existing adopters determines the utility for additional adopters (Fig. 5.1).

[1] The *direct utility* is also termed *autarky value* and is the value generated by the product, even if there are no other users. The *synchronization value*, in contrast, results from additional value

A. Kemper, *Valuation of Network Effects in Software Markets*,
Contributions to Management Science, DOI 10.1007/978-3-7908-2367-7_5,

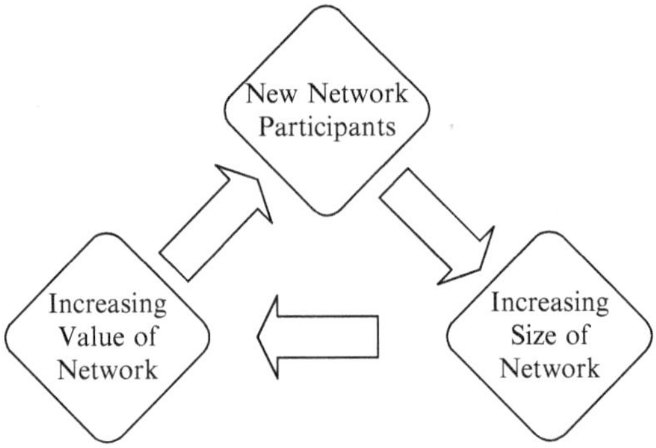

Fig. 5.1 Network effects in software markets
Source: Author

More formally, the overall utility U of a product i is the sum of its autarky utility a and its derivative utility b such that

$$U_i = a_i + b_i * N, \tag{5.1}$$

where N is the expected size of the customer network (Kotler and Bliemel 1999; Gareis 2000).[2] The relative intensity of network effects Q can be expressed as the ratio of the derivative network utility and the original utility, such that

$$Q = \frac{c}{c + b}, \tag{5.2}$$

with the original utility b and the derivative network effect utility c. Accordingly, the role of network effects increases with increasing Q (Buxmann 2001).

Research on network economics has a long tradition. Initial papers on the Veblen effect date back to the end of the 19th century.[3] Differences between the various effects become clear if static and dynamic effects are distinguished with respect to supply and demand. The resulting classification is provided in Table 5.1. Accordingly, the bandwagon effect and the economies of scale are static concepts, whereas the learning effects and the network effects are dynamic concepts (Chou and Shy 1990). Although outlined economic effects and their interactions can influence valuation in software markets, but the focus of this research is on network effects. Hence,

---

derived from being able to interact with other users of the product. Please confer Sect. 3.3 for additional definitions.

[2] In this notation the utilities are expressed as net values after the subtraction of the respective costs.

[3] *Veblen goods* are characterized by demonstrative consumption and status-seeking as the preference for them increases as a direct function of their price (Veblen 1899). This behavior reveals the presence of negative network effects. Examples of such high-status goods are diamonds, champagne and luxury cars.

Table 5.1 Classification of economic effects

Effects	Static effects	Dynamic effects
Demand-side effects	Bandwagon effect	Network effects
Supply-side effects	Economies of scale	Learning effects

Source: (Chou and Shy 1990)

the most relevant developments in network economics with a focus on network effects can be clustered in the following four phases.

1. *Bandwagon Effect.* In 1950, Harvey Liebenstein wrote a seminal paper about the bandwagon effect. This is an interaction of demand and preference that arises when the demand for a product increases in line with an increasing popularity (Mankiw 2004). In economic terms, the demand curves are more elastic if consumers derive an increasing utility from a growing market size. Despite this early progress, the findings remained unexplored for some decades (Rohlfs 1974). This is a contradiction to the conventional theory of supply and demand, which assumes that buying decisions are solely based on prices and personal preferences.
2. *Break-up of Natural Monopolies.* Network effects were rediscovered in the 1980s, when researchers studied the properties of telephone networks motivated by discussions concerning the break-up of "natural" monopolies, e.g. telephone networks (Katz and Shapiro 1985). Industrial economists investigated whether natural monopolies lead to inefficient market results, and whether governmental supervision is required in telecommunication markets. The discussion of competition in network effect markets launched a fierce discussion concerning demand interdependency and desired market outcomes.
3. *Economics of Standardization.* Another stream of literature emanated from the desire to understand standardization and technological innovation processes (David 1985; Farrell and Saloner 1986; Besen and Farrell 1994; Shapiro and Varian 1999). A famous example of a standardization war is the battle between VHS and Betamax (Katz and Shapiro 1986; David and Bunn 1988; Arthur 1989).
4. *Recent Developments.* Today, network economics is a systematic body of research on the characteristics, market structures and market performance of network effect markets, with considerable empirical and theoretical contributions (Tirole 1999; David and Greenstein 1990). Its main goal is to study the strategic implications of interdependent consumer decisions based on social interactions (Shy 2000b). Most recent research contributions study the relevance of network effects in diffusion of innovative technologies such as Voice over IP and fuel cell cars (Hensel and Wirsam 2008; Kellner 2008).

Network economic research reveals that software markets are strongly affected by network effects as the derivative utility of software can be significant. Hence, it contains a considerable explanatory potential to enhance our understanding of valuation in software markets.

5.2 Typologies of Network Effects in Software Markets

A literature review reveals that there are various classifications for network effects. But the following two are particularly relevant for valuation in software markets:

- Direct and Indirect Network Effects
- Application, User, and System Effects

Both classifications cluster the network effects with respect to different dimensions as depicted in Fig. 5.2. The difference between both concepts is explained in the following subsections.

5.2.1 Direct and Indirect Network Effects

Regarding compatibility network effects can be differentiated into direct horizontal effects resulting from an installed base, or as indirect vertical effects stemming from a network infrastructure. *Direct network effects* are generated through a direct physical effect from the number of purchasers, whereas *indirect network*

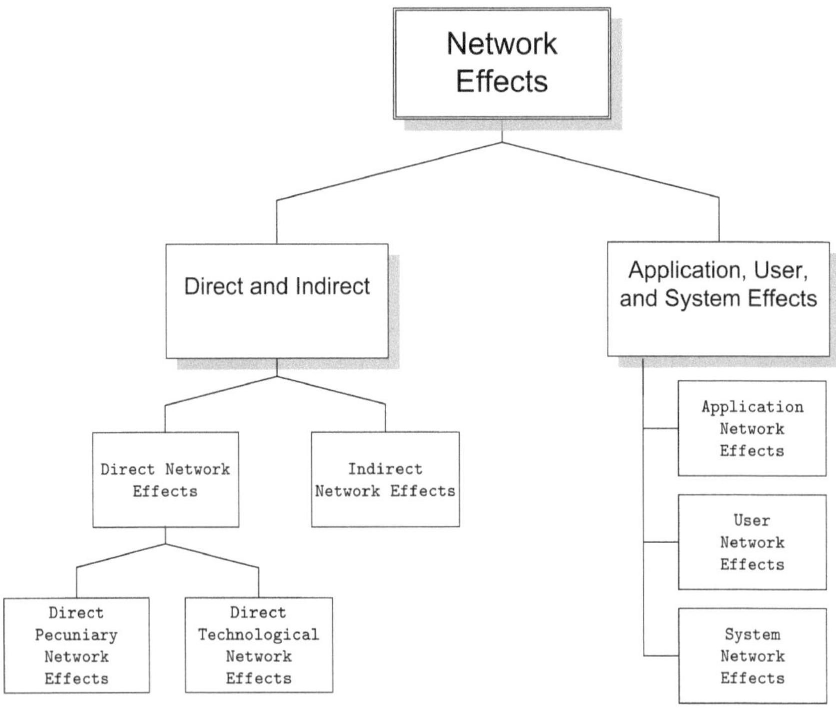

Fig. 5.2 Typology of network effects
Source: Author

effects are market mediated effects. In other words, horizontal networks are driven by customers, while vertical networks depend on complementary products (Ichbiah 1993). *Horizontal direct networks* are characterized by compatibility which increases the utility of the product with a growing number of installations, an example is the keyboard standard QWERTY (David 1985; Müller 1990). As the consequences of internalizing direct and indirect network effects are quite different, they are further subdivided into direct pecuniary effects influencing the price mechanism of the product, and direct technological effects increasing the utility of the consumers (Liebowitz and Margolis 1994). A few empirical papers study direct network effects, e.g. the investigation of the demand curve in the facsimile market (Economides and Himmelberg 1995). *Indirect network effects*, in turn, derive from complementary products which are beneficial for all network participants, e.g. PC and software (Farrell and Saloner 1985; Economides 1996). While products in horizontal networks increases the utility, a vertical complementary product only increases the demand. They are market mediated effects deriving from the benefits of complementary goods and services (Economides 1996). An excellent example in software markets is the hardware-software paradigm of Microsoft (Katz and Shapiro 1985). Accordingly, the operating system has a broad installed base that reinforces the development of applications, which in turn increase the popularity of the operating system (Arthur 1996). From an economic perspective, the increasing number of network participants results in an increasing return to scale, which allows for the development of additional competitive advantages from lower prices or investments in higher quality (Achi et al. 1995). Indirect network effects can be assessed by simultaneously estimating the demand for multiple components, and testing their interdependence, e.g. compact discs and VCRs (Gandal 1994; Ohashi 2003). In essence, horizontal direct networks are a special class of vertical networks, in which the installed base product and the network product are the same. Combinations of both effects are commonly classified according to their more predominant attribute, e.g. spreadsheet programs that depend on the installed hardware as well as on compatible operating system (Groehn 1996). Interdependent networks are special cases as they require an infrastructure for other products, e.g. credit card networks that depend on customers and on acceptance authorities.

5.2.2 Application, User, and System Effects

Alternatively, network effects can be clustered along the various classes of software into application effects, user effects, and system effects (Groehn 1999). While *user effects* are based on an increasing number of additional users, *application effects* derive from the utility to use compatible applications in growing networks. *System effects*, in contrast, result from the interdependency between software and hardware. This classification is mentioned as all three types of network effects can be relevant in software markets, depending on the business model and the business segment of the investigated company.

5.3 Properties of Network Effects in Software Markets

Network effects markets differ from traditional markets with respect to a variety of economic properties. Investigations emphasize topics such as the startup problem, market failure, instability of network markets, and path dependency (Rohlfs 1974; David 1985; Farrell and Saloner 1985; Arthur 1989; Wiese 1990; Besen and Farrell 1994; Katz and Shapiro 1994; Economides and Himmelberg 1995; Arthur 1996; Weitzel et al. 2000). An overview of the review is provided in Table 5.2, based on which the most relevant properties are investigated in the following section. Accordingly, the following properties of network effects markets are particularly relevant to valuation in software markets. The most important properties of network effects are investigated in detail in the following subsections.

5.3.1 *Feedback Loops*

Network effects markets are characterized by positive and negative *feedback loops* (Katz and Shapiro 1994). Feedback is a cybernetic mechanism, in which some output of a system is looped back to its input (Sterman 2000). Consequently, feedback and control are self-related. While *positive feedback* amplifies signals and allows a system to access new points of equilibrium, *negative feedback* loops help to maintain stability in a system despite external changes. *Bipolar feedback* loops either increase or decrease the output of a system. The concept of negative feedback was first discovered in electrical amplifiers in 1927, but it became popular in 1943 when Arturo Rosenblueth set the basis for cybernetics and control theory by proposing that feedback is a determinative phenomenon in nature and human creations. (Rosenblueth et al. 1943) Today, such feedback loops are identified in most complex systems in engineering, architecture, economics, and biology (Bar-Yam 1997). In software markets, feedback loops are relevant as the autoregressive expectations of customers shape the competitive landscape. Empirical studies illustrate that consumers frequently do not base their decisions on rational expectations but rather tend to follow trends based on imperfect information (Besen and Farrell 1994; Regibeau and Rockett 1996). This relationship underlines the importance of information management and investor relations in software markets. Announcements of product launches or of compatibility influence the purchasing decision of customers and, thereby, the future development of the company (Swann and Shurmer 1994). Hence, the reputation of the software company is a decisive factor for the success of a software which can be supported by credible up-front investments that have an influence on the autoregressive expectations of consumers, banks and financial investors (Katz and Shapiro 1994; Graumann 1993).[4]

[4] Please note the vital implications of this aspect for turnaround strategies of software companies as depicted in part 16.1.3.

Table 5.2 Analytical studies on network effects

Study	Goal	Approach	Installed base
Rohlfs (1974)	Description of a cumulative demand function	Equilibrium analysis	$q_{i^D} = q_i^{D(t-1)}$, $(p, q_1, \ldots, q_{i-1}, q_{i+1}, \ldots, q_n)$, with $q_i = 1$ for network participants
Oren/Smith/ Wilson (1982)	Optimal tariff policy in communication networks		$w(q, t, Y)$, with $W(q, t)$ as original utility for first unit q of t consumer types and network effect Y as derivative utility
Katz/ Shapiro (1985)	Welfare analysis	Equilibrium analysis	$U_i = r + v(y_i^e) - p_i$, with original utility r and derivative utility $v(y_i^e)$
Dhebar/ Oren (1985, 1986)	Monopolistic Intertemporal Pricing		$W(\eta, X)$ or $W(q, \eta, X)$, where the utility of the individual η at network size X is determined, while q is a quantity volume.
Farrell/ Saloner (1986)	Excess inertia and excess momentum	Equilibrium analysis	$u(x) = a + bx$, with original utility a and derivative utility u(x)
Arthur (1989)	Path dependency	Equilibrium analysis	$a_R + rn_A, b_R + rn_B$, with actor R preferring technology A and actor s preferring technology B
Wiese (1990)	Price strategy	Duopoly simulation model	$U_i = U_i(z_i, x_i, l)$, with utility of individual i from consumption of z_i units of network x and the total number l of consumers x
Church/ Gandal (1996)	Indirect network effects	Equilibrium and welfare analysis	interdependency of software and hardware

Source: Author based on (Weitzel et al. 2000) and (von Westarp 2003)

5.3.2 Hysteresis, Startup Problem and Critical Mass

The outlined feedback loops cause a characteristic path dependent development of software markets, which is also termed *hysteresis effect* (Clement et al. 1998). This effect states that the product is only interesting for potential customers if a critical mass of consumers is reached such that the sum of original and derivative utility outweighs the respective costs (Economides and Himmelberg 1995; Clement et al. 1998; Choi and Winston 2000; Wamser 2000; Wirtz and Kam 2001). As the popularity of a software product can vitally depend on the expectations of customers, its success can be a self-fulfilling prophecy that implies the iterative problem to determine the critical mass of customers (Katz and Shapiro 1985). The outlined hysteresis effect can cause a *Startup problem* during the market introduction of software products (Wiese 1991). The dilemma is that software with a small customer network has considerable problems if its value is primarily based on a derivative utility. Early adopters bear the risk to strand with their investment and risk switching costs if the purchased technology is not supported by a critical mass of customers. This problem becomes even more important if market entry barriers prevent efficient competition in software markets and if the company is itself a Startup.[5] Hence, a *critical mass* of customers is required, which has a crucial role in the outlined network effects. But there are only vague approaches to quantify it in software markets.[6] According to a stream of research, the critical mass consists of 10–20% of the total population (Rogers 1995). This assumption is based on the underlying hypothesis that technological adopters are normally distributed, such that the sum of innovators and early adopters is sufficiently above the technology adoption gap.[7] In contrast, a second stream of research states that the critical mass depends on the nature of the product and its respective derivative utility (Clement et al. 1998). Finally, a third school of thought does not assume an absolute number but more generally some communities of interest, which all influence the buying decision of customers. The quantification of the critical mass in customer networks of companies operating in software markets is a central research objective that determines the following parts.

5.3.3 Excess Momentum

Network effects can also result in *excess momentum* if the software acquires a high market share within a very short time due to a self-containing propagation process (Bower and Christensen 1995). Empirical network effects research indicates that such extreme tipping dynamics can imply that the winner takes the whole market (Liebowitz and Margolis 1990). Popular examples of such volatile developments in software markets are WordPerfect and Lotus 1-2-3. In summary, it is important to note that few dynamics in the form of the startup problem and excessive dynamics

---

[5] Please confer Sect. 3.3.10.

[6] Please note that this critical mass of customers can be different from the number of customers required to reach a break-even calculation, i.e. if revenues cover the costs and profits are zero.

[7] Please confer also Sect. 7.2.1.

in the form of excess momentum have the potential to delay or even to prohibit the diffusion of software (Besen and Farrell 1994).

5.3.4 Market Instability and Serial Monopolies

The stability of software market developments is the subject of another discourse in economics. While some streams of research assume that network effects induce a permanent propensity to monopolize, more recent research emphasizes their finite character in the form of a sequence of *serial monopolies* (Liebowitz and Margolis 1999, p. 10).[8] Nevertheless, research confirmed that software markets frequently have a propensity to monopolize because of multiple, incompatible technologies that all compete for a dominant market position. They are instable as rivaling technologies, they do not coexist and the high dynamics in the standard setting process do not provide much time for strategic reactions (Besen and Farrell 1994). Once a standard is set, the coexistence of several products in multiple segments depends on the market size and on the intensity of the competition. Empirical studies indicate monopolies due to market instabilities in the market for office suites, word processors, and spreadsheet software (Groehn 1999; Liebowitz and Margolis 1999).

5.3.5 Compatibility

Compatibility between products in software markets is a prerequisite for network effects (Katz and Shapiro 1985; Wiese 1990). It is defined as the possibility of a reciprocal exchange of goods or data among various systems (Farrell and Saloner 1987). Based on this definition, three classes of compatibility are distinguished:

1. *Physical Compatibility*. Physical compatibility requires that an interface is a physical part of the product.
2. *Communication Compatibility*. Communication compatibility implies a communication interface between two products, such as protocols in computer networks.
3. *Contractual Convention*. Contractual conventions are interfaces between two systems that are defined in a contract.

All three classes can be further distinguished into one-sided and two-sided compatibility (Herget 1987). While two-sided networks are characterized by complementary parts that have a greater value if they are combined, there is no additional combinatoric benefit in one-sided networks (Economides and White 1993).

[8] A serial monopoly is a monopoly or a sequence of near monopolies.

5.3.6 Multiple Market Equilibria

Research in industrial economics reveals that the outlined characteristics of network effects markets give rise to a peculiar competitive behavior and market outcome (Topkis 1979; Salop and Scheffman 1983; Bulow et al. 1985; Milgrom and Roberts 1990; Armstrong 1998; Laffont et al. 1998a; Laffont et al. 1998b; Carter and Wright 1999; Vives 1999; Dessein 2001; Laffont and Tirole 2000). *Multiple market equilibria* are possible as the transaction costs and the equilibrium set of users are frequently very sensitive to changes of the input parameters (Rohlfs 1974).

5.3.7 Multiple Product Generations

Network effects of installed bases from different product generations can interact with each other if they are compatible. As a consequence, new products can benefit from existing customer networks that allow them to instantaneously internalize the derivative value. Compatible networks reinforce each other, as users of the old generation have the opportunity to switch to the more innovative version. If, in turn, products are not downwardly compatible there is a trade-off between compatibility and innovation which is likely to delay the innovative product (Farrell and Klemperer 2001). Such intergeneration network effects have to be considered for valuation in software markets as they can influence the profitability of investments, i.e. network effects of platform investments that create managerial flexibility.

5.3.8 Pareto-Inferior Market Results

The network effects are subject to an extensive controversial discussion as some researchers claim that they infer *pareto-inferior market results*, in contrast to other authors, who assert that network effects facilitate the diffusion of superior products (David 1985; Arthur 1990; Liebowitz and Margolis 1994). In essence, this discourse is based on the following views:

1. *Loss of Variety*. A standardization process is positive for consumers due to gross consumption benefits. But at the same time such effects can result in network congestions or in a loss of variety of suppliers (Farrell and Saloner 1986). Moreover, increasing concentration of purchasing power is assumed to be negative for fragmented consumers, as they cannot benefit from consumer sovereignty (Besen and Johnson 1986).
2. *Delay of Innovations*. Another reason for inefficient market results is that excess inertia favors old technologies, whereas strong network dynamics speed up a product adoption process. Marketing strategies supporting only one standard reinforce a convergence toward one standard product, as the initially high variance decreases over time. Free market entries are only possible if the demand is

elastic enough. This implies that dominant corporations are frequently capable of inhibiting competition (Köster 1998).

3. *Ex-post versus Ex-ante standardization.* A further reason for inefficient market developments is the trade-off between standardization by experiment, i.e. ex-post standardization, and standardization by coordination, i.e. ex-ante standardization (Choi 1996). Ex-ante standardization provides additional utility, but as the quality of the product is unknown, it is possible that results below the social optimum occur, if the loss due to the reduced quality is higher than the additional gains by standardization.
4. *Tragedy of the Common Problem.* Furthermore, although networks reveal positive effects on an aggregated level, but individuals consider only their personal utility. Consequently, decentral decision making can lead to overall pareto-inferior market results (Arthur 1989).

With respect to valuation in software markets it is important to note that network effects increase the complexity of possible market developments. This makes it necessary to investigate the development of the software market in detail based on simulations of a software market model. All of the outlined network market dynamics have in common that they influence the business development. Growth induces growth. Thereby, the identified feedback loops influence valuations in software markets and determine whether a company is successful or fails.

5.3.9 Competition For and In Software Markets

As both, the direct and the strategic network effects, can be market entry and exit barriers each must be considered, if the competitive dynamics of software markets are investigated (Wied-Nebbeling 2003). The exact influence depends on the compatibility of the software and the respective size of the customer network. Software firms *compete for markets* against each other with incompatible software and *in markets* with compatible software in order to establish a market (Besen and Farrell 1994). In small networks network effects increase the concentration of suppliers as incompatibility is a competitive advantage of in-market software companies (Matutes and Regibeau 1996). Compatibility in large networks, in turn, lowers the opportunity for product differentiation and is, hence, an incentive for a market entry (Matutes and Regibeau 1988). Rents of first-movers attract imitators, but they can be welcome to first-movers if they support the diffusion of a new technology, as the overall increasing demand may compensate the lower prices due to competition (Economides 1996). Thus, competing software companies may cooperate in order to jointly develop an emerging market with a common standard, which otherwise would remain untapped. But once a standard emerges, the coexistence of rivaling products depends on the market size and on the degree of competition. A reason for such cooptition is the high stranding costs for consumers as they fear to make buying decisions and may prefer to wait until a standard emerges. This behavior is also known as the *Penguin effect*.[9] In this cooptition credibility and trust are vital success factors.

5.4 Empirical Evidence of Network Effects in Software Markets

Empirical research in network economics underlines the importance of network effects for understanding the dynamics of software markets. A very popular class of empirical studies investigates network effects based on a regression of a hedonic price function (Hartmann and Teece 1990; Gandal 1994; Economides and Himmelberg 1995; Brynjolfsson 1996; Groehn 1996; Groehn 1999). This method is applied in economic research to estimate values that directly affect market prices (Anglin and Gencay 1996). The basic premise of the hedonic pricing method in order to investigate the network effects is that people value the characteristics of a good, or the services it provides, rather than the good itself. For example, the price of a car reflects some of its characteristics such as transportation, comfort, style, luxury, etc. Therefore, it is possible to value the individual characteristics of the good by looking at how the prices that people are willing to pay for it change if the characteristics change. Consequently, prices are interpreted as the value of a set of characteristics that people consider to be important if they purchase the product. This hedonic pricing method is frequently applied, as it is based on actual market prices and fairly easily measured data. For example, empirical investigations of the sample period from 1985 to 1995 reveal the importance of network effects in word-processing software using a hedonic price approach (Groehn 1999). The corresponding regressions explain up to 74% of the price variations in the market, despite the considerable variance and dynamics of the software industry. Therefore, the study concludes that network effects have a significant impact on prices in the word processor segment (Groehn 1999). Additional supportive studies with similar findings investigate network effects in German and US software markets. They conclude that products are similar in both countries in the segments of email, office communication, network protocols, database systems, programming languages and operating systems, whereas business software segments are different (von Westarp 2003). While two thirds of the investigated German corporations apply one or two products, only half of the US corporations are limited to so few products, while the other half uses three or more products (von Westarp 2003). This difference can be explained by the dominant position of SAP in Germany which benefits from the respective network effects in the customer network. Among all segments the

[9] In this analogy, a group of Penguins is imagined to gather on an ice floe and while all are hungry nobody wants to be the first to jump into the water. But as soon as the first Penguin jumps into the water all companions follow the leader.

Table 5.3 Empirical studies on network effects in software markets

Study	Market	Goal	Approach
Hartmann/ Teece (1990)	Hardware	Significance of network effects	Regression of hedonic price function
Gandal (1994)	Spreadsheet software	Significance of network effects	Regression of hedonic price function
Moch (1995)	Database software	Significance of network effects	Regression of hedonic price function
Gröhn (1999)	Word processing	Significance of network effects	Regression of hedonic price function
Gallaugher/ Wang (1999)	Web Server	Correlation of price and market share	Regression of hedonic price function
Westarp (2003)	ERP, EDI and Office Suites	Corporate adoption behavior	Survey, simulations and case studies

Source: Author based on (Groehn 1999) and (von Westarp 2003)

programming and operating systems have the largest variety. Only 30% use less than three programming languages, and even fewer corporations use less than three operating systems (von Westarp 2003). In turn, only a few companies operate with more than one email program. Up to 65% use only one email software solution, while only 8% of the German and 14% of the American corporations use three or more email programs. With respect to compatibility, corporations operating in the US fear, more than their German colleagues, to encountering an incompatibility problem (von Westarp 2003). In the business segment, the fear concerning incompatibility is highest, which is probably resulting from the significant strategic importance of such programs to the corporations as it effects all key processes. Empirical research reveals that the statistical correlation is highest in the business software segment. In summary, it is reasonable to conclude that software incompatibilities are an important aspect of corporate decision making in software markets and underline the vital importance of network effects (Table 5.3).

The outlined empirical research confirms the hypothesis that network effects have a vital impact on the dynamics of software markets. But at the same time, the empirical review also indicates that some segments are more affected than others by network effects. Therefore, selected sectors of the software markets are investigated on a higher level of detail in the following section.

5.5 Network Effects in Selected Software Market Segments

Empirical review revealed that the relevance of network effects is different across various segments of software markets. Hence, it is necessary to distinguish the various effects.[10] Software markets comprise, but are not limited to, the following relevant subsectors:

- Enterprise Resource Planning
- Electronic Data Interchange
- Office Suites
- Social Network Services

Although there are a variety of additional subsectors, such as computer entertainment or security software, a comparison of the selected segments reveals the broad variety of network effects in software markets.

5.5.1 Network Effects in Enterprise Resource Planning

Enterprise resource planning (ERP) mainly support large organizations by integrating all data and processes into a unified system (Monk and Wagner 2006).[11] They are based on a unified database that stores data for modules that control many business activities. Examples are financing, manufacturing, logistics, distribution, inventory, shipping, invoicing, and human resource management. The market for ERP systems underwent a very volatile development. The demand for ERP programs has grown significantly since the end of the 1980s. The main influences were that logistics, sourcing and capacity plannning became major parts of the standard software architecture. In the 1990s, ERP markets saw a large boost in sales due to the millennium problem. But the rapid growth in sales stopped in 1999 when most companies had already implemented a solution. A comparison of the German and the US markets reveals that the underlying customer networks are driven by different dynamics. In the US, the supply side is heterogeneous, but shares a similar market share, while SAP dominates the German market. An empirical comparison of the price elasticity of German and American managers reveals that German MIS managers are more price sensitive than their US counterparts. Network effects are comparatively low in the ERP market, as the software is frequently bought by a central authority in order to avoid incompatibility. Such problems are a major concern for MIS managers who are responsible for the IT infrastructure. The ERP software of business partners is relatively irrelevant, as the exchange of sensitive internal data is relatively rare. In addition a variety of interfaces are available in order to exchange data among various ERP systems. Therefore, purchasing decisions are more likely based

---

[10] Please confer Sect. 5.4 for the empirical review.

[11] The term originally derived from manufacturing resource planning (MRP II) that followed material requirements planning (MRP).

on the functionality of the software, since customization and Inhouse developments are very expensive. Nevertheless, empirical research indicates that a large majority of decision makers in the US and in Germany are convinced that standardized business software becomes increasingly important (von Westarp 2003). This implies that network effects are likely to become also more important.

5.5.2 Network Effects in Electronic Data Interchange

Electronic data interchange describes an inter-company, application-to-application communication approach to electronically exchange data from business transactions in standard formats within businesses, organizations, governmental institutions or other groups. The underlying idea is a transfer of structured data, based on agreed standards, from one computer system to another without human intervention. The term is also frequently used in order to refer to the implementation and operation of systems and processes for creating, transmitting, and receiving EDI documents. Despite other technological innovations, such as XML web services, the Internet and the World Wide Web, EDI is still a frequently used standard in electronic commerce transactions around the world. The market demand is dominated by large companies who benefit most from the economies of scale resulting from the cost savings, but upcoming standardized Internet-solutions will make the products more attractive to Small and Medium Enterprises. Due to various market conditions and different customer networks, the suppliers of EDI software tend to specialize in industries. The German EDI market is rather heterogeneous, while most products from the US market are compatible. While custom-made EDI solutions are currently in a high price segment, the emergence of standardized WebEDI solutions indicates a trend towards lower price segments. On the contrary, compatibility with solutions for external business partners is a main concern driving the decision making of MIS managers, while functionality and price are less important (von Westarp 2003). In addition to the network effects by the standard which determine the contemporary decision making, the network effects of supplier related standards will become more important with the increasing emergence of standardized supplier solutions. While EDI solutions were mainly based on the standards and the custom-made solutions, the emerging products with flexible data formats will allow more standardized solutions, that increase the network effects of suppliers.

5.5.3 Network Effects in Office Suites

An *office suite* is a software package intended to support typical clerical worker and knowledge workers. Its goal is to support the design, presentation and publishing of office documents, e.g. word processing, spreadsheets, database, and presentation tools. Particular characteristics of this type of software are that the individual components are distributed together, have a consistent user interface and usually can interact with each other. From a supply-side perspective, the most popular office suite is Microsoft Office, which has become a proprietary de-facto standard in office software. Alternatives are OpenDocument suites, such as OpenOffice.org or StarOffice. Moreover, innovative online word processors and office solutions, such as those offered by Google, allow one to centrally edit stored documents by using a web browser. The demand for office suites, in turn, is heterogeneous as it consists of companies and personal users. In contrast, the supply side is characterized by a worldwide monopolistic structure dominated by Microsoft. This monopolistic market position induced antitrust litigation which is a popular topic of network economic controversial discussions (Liebowitz and Margolis 1999). The focus of the discussions is the practice of software suppliers to bundle integrated services. This limits competition as economies of scale allow the placement of such products in the low price segment of network markets (von Westarp 2003). Low prices enable users to switch between various applications (Liebowitz and Margolis 1999). While some researcher state that network effects and compatibility are important for the exchange of data, whereas other researchers are convinced that superior product quality is the decisive factor in competition (Liebowitz and Margolis 1999).

5.5.4 Network Effects in Social Network Services

Social network services focus on the setup and maintenance of social networks for communities sharing common interests and activities, based on the use of online software solutions. Such applications connect people at low costs, which can be beneficial for private as well as for business purposes. The services are mainly based on web solutions and provide various interaction channels for users of the network, such as chat, messaging, email, video, voice chat, file sharing, blogging, and discussion groups. As such services are globally available, social networks allow to keep in touch with contacts around the world. Within a broad spectrum of social networking services, nearly all platforms contain directories, connection options to acquaintances, and recommender systems or a combination of the outlined features. The number of platforms increases rapidly, but the most popular examples are MySpace, Facebook, Bebo, GalaOnline in anglo-saxon regions, Orkut and HI5 in Latin America, Friendster and CyWorld in Asia, and Bebo, MySpace and Xing AG in Germany (Fulgoni 2007). Network effects are the main driver of their business model. Participants are willing to pay for such software solutions if the benefits outweigh the costs. While the original value of a social network site is considerably low, e.g. to store the personal information, the derivative utility resulting from a large customer network represents a significant percentage of the total benefit. Therefore, social network services are illustrative examples of business models in software markets which are based on network effects, i.e. demand-side economies of scale.

5.5.5 Comparison of Software Market Segments

A comparison of the selective segments reveals that network effects are specific to various software market segments. Direct network effects are less important factors for ERP software, whereas they play a decisive role in EDI markets. In turn, it is important to note that the individual network of business partners is important in some segments, while the total installed base is the decisive network in others. Thus, it is necessary to design accurate market models for valuations in software markets which account for the outlined features. Moreover, the comparison indicates that social network service platforms are illustrative examples of software segments that are vitally effected by network effects.

5.6 Reconsideration of Valuation in Modern Software Markets

The investigations of network effects in software markets provide a variety of insights relevant to valuations in modern software markets. First, software markets are network effect markets. Hence, it was necessary to investigate the principles of network economics in order to increase understanding of the dynamics of software markets. In a further step the most important properties of software markets relevant to valuations were outlined. Empirical research supports the particular relevance of network effects for software markets and reveals that various segments of the software market are characterized by different types and intensities of network effects. Despite such contributions it is also possible to consider some limitations.

Perfect Information and Complete Rationality. New institutional economics suggest that perfect information and rationality of all market participants are questionable assumptions. (Hodgson 1993) In the real world, parametric and strategic uncertainty as well as heterogeneous institutional and structural influences create constitutional bounds to the possible level of information (Hayek 1937). Similarly, complete rationality is rejected and replaced by a model of bounded rationality (Hodgson 1993).

Consumption Paradigm. Classical models suggest that the utility of a product is derived by its consumption, but information goods such as software are used and not consumed. Therefore, usufruct rights become important and the perspective changes from consumer to user. Similarly, economic agents are traditionally classified into producers and consumers, ignoring the increasing importance of prosumers who consume and produce simultaneously.

Divisibility of Goods. The divisibility of goods implies that all goods can be divided into parts that can be sold separately. In software markets a single product is frequently the optimal quantity as multiple copies of a software usually do not provide additional utility.

Special Cost Structure. A questionable assumption of the network effect theory is that network effects constantly increase due to hypothetical minimal marginal costs, and as network entry costs are ignored. This implies the hypothesis of constant, or even falling, marginal costs for new members. This is also questionable as such simplifying assumptions do not apply to all industries, as this implies an inexhaustible economy of large-scale operations (Liebowitz and Margolis 1994). Due to an increasing relevance of capacity constraints it is reasonable to assume that at a specific size of a network new members do not contribute further value due to increasing congestion costs (Matutes and Regibeau 1992). Moreover, transaction costs are not taken into consideration. Hence, small user islands can exist and multiple products can coexist.

Propensity to Monopolization. While various structures can be observed in real-world markets, network effect theory suggests that the competition in network effects markets ends either in a narrow Polypoly or in a Monopoly. Network economic research, however, is incapable of explaining other market structures, stable user groups nor the temporal coexistence of products in software markets, e.g. the coexistence of Microsoft Windows, Linux and Mac OS. Consequently, it is necessary to extend the existing models in order to explain the variety of possible outcomes that can be observed in real world networks. The installed base is a switching barrier for superior new technologies as collective switching from the legacy network is difficult due to the coordination problem (Farrell and Saloner 1985; Katz and Shapiro 1986; Arthur 1989). Recent research studies challenge the single perspective on the welfare implications (Liebowitz and Margolis 1999). Software specific approaches argue that not all network effects are externalities as vendors can internalize at least a part of the network effects with property rights. Particularly, if several products compete in software markets, they are not required to fail but to innovate (Shapiro and Varian 1998; Liebowitz and Margolis 1999). Therefore, microeconomic models, as well as other areas of management such as finance, have to be reconsidered.

In summary, the design of software market models for valuations has to particularly account for the following issues.

Network Effects. The model has to account for the diffusion dynamics of innovative products in software markets.

Bounded Rationality. Network participants have limited perception capabilities.

No convexity and divisibility. As the consumer choices are rather discrete and interdependent, the convexity and divisibility assumptions of the neoclassical model are inappropriate.

The outlined network economic perspective illustrates that the existing approaches are neither capable to explain nor to model the diffusion dynamics of software markets. Existing network economic models are reasonable approximations for markets with moderate consumer interdependency and adoption rates. But since software markets are different, further research is required. After network effects have been identified as relevant factors, their description, quantification and analysis of the network effects are the research goals of the following research. For this purpose, research on customer-equity valuation, as well as on adoption and diffusion research, is reviewed in the following chapters.

Chapter 6
Customer-Equity Valuation in Software Markets

> "What's a customer worth? The company that can answer this question precisely is the company with an edge in the customer-based, technology- and information-intensive economy of today. But how can an asset as intangible as customer value be measured?"
>
> (Blattberg et al. 2001)

The previous investigations revealed the importance of network effects for valuations in software markets. In the next step the findings on network effects are linked to valuations in software markets based on a customer-centric valuation approach. First, the background of customer-based valuation is presented. Then, a DCF customer equity model is outlined before a real options version is reviewed. As both models reveal the importance of reliable software market models based on solid adoption and diffusion models, these are investigated in the next chapter. Respective insights provide the basis for the development of a network effects framework for valuations in software markets at the end of this part.

6.1 Principles of Customer Equity Valuation

A *customer network-centric valuation approach* is an innovative perspective on valuation in software markets as it emphasizes the customer base, i.e. the customer network, as a vital value driver of companies operating in software markets. Traditional models consider primarily initial product purchases, but research on network effects in software markets reveals that it is also necessary to consider the revenues resulting from the interdependencies among customers in customer networks.[1] Following this reasoning, the customer lifetime value (CLV) approach considers all cash flows during the entire customer life cycle (Blattberg and Deighton 1996; Blattberg et al. 2001, Staat et al. 2002). Based on principles of contemporary finance

[1] Research differentiates preacceptance, acceptance, and post-acceptance during a purchasing process of a product (Kollmann 1998). While the pre-acceptance phase describes the expectations of users before the purchase of a product, the post-acceptance phase refers to the time after the purchase.

A. Kemper, *Valuation of Network Effects in Software Markets*,
Contributions to Management Science, DOI 10.1007/978-3-7908-2367-7_6,

its goal is to determine a profitability metrics of customer relations including profits resulting from network effects (Day and Fahey 1988; Doyle 2000). It is quantified as the net present value of the customer relationships based on customer retention and migration models (Cornelsen 2002).[2] This perspective integrates customer lifetime value and corporate valuation by breaking down the cash flows into individual components (Hoekstra and Huizingh 1999; Srivastava et al. 1999; Payne et al. 2001; Jain and Singh 2002; Bauer and Hammerschmidt 2005). The most relevant factors are revenues, costs, and retention rates. These are weighted with respective probabilities and discounted at a risk-adjusted rate (Reinartz and Kumar 2000). Despite considerable progress in research, some of the outlined variables are frequently not considered. While some models ignore customer retention rates (Jackson 1992; Mulhern 1999; Niraj et al. 2001), other approaches ignore reference revenues (Berger and Nasr 1998; Blattberg and Deighton 1996; Dwyer 1997; Blattberg et al. 2001; Wang and Spiegel 1994). Only a few models integrate multiple aspects into a single model. After an overview of the relevant parameters, the two most relevant customer equity models are outlined for valuations in software markets. While the first is an extension of the classical DCF approach, the second is based on the outlined real options perspective.

6.1.1 Revenue Parameter in CLV Models

Revenues can be split with respect to their sources into four categories, i.e. autonomous revenues, upselling revenues, cross-selling revenues, and reference revenues (Bauer and Hammerschmidt 2005; Kollmann 1998).

Autonomous Revenues. Autonomous revenues are generated by generic sales of a product. They are frequently approximated by time series or stochastic brand choice models (Lilien et al. 1992; Schmittlein and Peterson 1994). In software markets they are a large and volatile source of revenues.

Upselling Revenues. Upselling revenues are additional sales to loyal customers motivated by discounts, quantity effects or price effects which are frequently calculated with frontier function models that approximate the maximum revenues per customer relationship (Kim and Kim 1999).[3] In software markets, upselling potential is relevant due to software updates, whereas only a few customers buy multiple copies of a software.

Cross-selling Revenues. Cross-selling revenues, in turn, is the sale of complementary products which can be approximated by cross-selling matrices based on

[2] While *customer retention models* allow the derivation of retention rates calculated from historic cohorts of customers, *migration models* approximate purchasing probabilities based on historic data (Dwyer 1997).

[3] The price effect describes the sale of more expensive substitutes to long-term, less price sensitive customers, while the quantity effect derives from a higher transaction volume and frequency per period (Reichheld 1999).

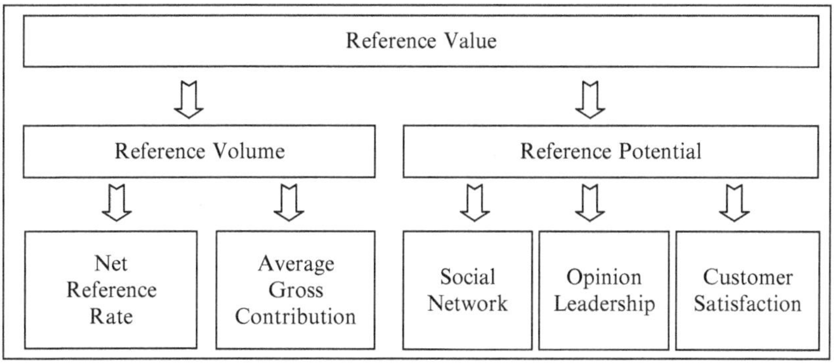

Fig. 6.1 Driver of reference value
Source: Author based on (Bauer and Hammerschmidt 2005)

experimental research, e.g. delivered by comScore (Reichheld and Sasser 1990; comScore 2008). In software markets, cross-selling potential can be significant due to network effects.

Reference Revenues. Reference revenues are generated with new customers based on referrals of existing customers due to network effects. The reference margin is the annual contribution of an average customer weighted by the degree to which references influence the purchase decisions of consumers. In this process, the reference potential is the ability of a customer to influence other customers (Cornelsen 2002).

While the first three streams are *direct sources* of revenues, reference revenues are *indirect sources* resulting from network effects. Their value contribution is also coined *reference value* (Bauer and Hammerschmidt 2005) (Fig. 6.1). The reference value is composed by reference volume and the reference potential. While the reference volume is determined by the net reference rate and the average gross contribution per customer, the reference potential is a derivative of the social network, opinion leadership and the satisfaction level of the customers.

6.1.2 Cost Parameter in CLV Models

Costs can be split into acquisition costs, marketing costs, sales costs and termination costs (Wang and Spiegel 1994). Fixed costs are excluded as these are not specific to the customers.

Acquisition Costs. Customer acquisition costs derive from marketing activities which aim to attract new customers and depend on the respective marketing strategy. Some segments of software markets are characterized by considerable acquisition efforts, such as ERP, in contrast to other segments, such as browsers, where the respective activities are relatively insignificant.

Marketing Costs. In a narrow sense marketing costs have a focus on customer retention and customer relationship in order to influence the upselling or cross-selling potential. Churn costs result from the loss of customers. Regaining costs are related to marketing efforts to win a customer back. In software markets the size of the marketing budget depends on the software market segment and the respective marketing strategy, e.g. costs are significant in the computer game sector due to aggressive marketing strategies in the growing profitable business (EITO 2006; comScore 2008).

Sales and Distribution Costs. Sales costs are primarily related to reproduction and logistics, which are frequently marginal due to download options.

Termination costs. Termination costs occur if customer relationships end. In software markets, such costs are frequently insignificant.

All costs have to be taken into consideration. The majority depends on the respective marketing strategy of the software company, which in turn is frequently related to the software market segment.

6.1.3 Retention Rate Parameter in CLV Models

Retention rates express the probability that a customer remains loyal for a certain period (Dwyer 1997). They can be quantified with causal analyses based on determinants such as satisfaction level, switching barriers, variety-seeking behavior, and competitors (Jones and Sasser 1995). Research reveals that two types of customer behaviors can frequently be identified. If switching costs are high, defecting customers are considered to be lost without a chance to regain them (Berger and Udell 1998). Alternatively, if consumers are very flexible, the switching behavior can be described based on Markov-Chain models (Dwyer 1997; Schmittlein et al. 1987).

6.2 DCF Customer Equity Model

The customer equity approach is different from the classical DCF approach as it disaggregates cash flows on the level of individual customers. Main input variables are the growth of sales, the return on sales, the income tax rate, the investments, the cost of capital and the value growth duration, with a focus on the first two factors. Various revenue streams are differentiated based on the outlined typology and all current, as well as potential customers, in the customer network are taken into consideration. Accordingly, the overall corporate value is defined as the sum of the customer-based operating cash flows and of the non-operating activities (Gupta et al. 2001).

6.2.1 Generic Customer Lifetime Value

Based on the outlined terminology, the customer lifetime value of a consumer is defined as

$$CLV_i = -AC_i + \sum_t^T \left(r_i^t \frac{(AR_{ti} + UR_{ti} + CR_{ti} + RV_{ti}) - (SC_{ti} + MC_{ti})}{(1+d)^t} - (r_i^{t-1}(1-r_i)) \frac{TC_i}{(1+d)^t} \right),$$

with the net present lifetime profit CLV_i, the acquisition costs AC_i, the retention rate r_i, the autonomous revenues AR_{ti}, the upselling revenues UR_{ij}, the crosselling revenues CR_{ij}, the reference value RV_{ij}, the marketing costs MC_{ti}, the cost of sales SC_{ti}, the termination costs TC_i, the risk-adjusted discount rate d and the projection period T (Bauer and Hammerschmidt 2005). The probability of autonomous migration can be quantified with stochastic choice and attraction models (Bayon et al. 2002). The infection rate for new customers is inversely related to the retention rate and can be determined by considering the expenditures, the constant acquisition efficiency as well as the upper acquisition threshold (Wang and Spiegel 1994; Blattberg and Deighton 1996). The probability that customers remain loyal is frequently approximated by empirical or historical data (Schmittlein and Peterson 1994).

6.2.2 Individual Customer Lifetime Value

The lifetime value of an individual customer can be approximated by

$$CLV_{i0} = \sum_{t=0}^T r_i^t \frac{(R_{ti} - C_{ti})}{(1+d)^t}. \tag{6.1}$$

Accordingly, new cohorts emerge in every period s and their profit margin $R - C$ is likely to vary over time. The sign of the variations depends on the individual situation. While some empirical studies indicate accelerating profits because of increasing purchase frequencies, decreasing costs, increasing switching costs, and increasing network benefits, other researchers conclude that long-term customer relationships are not profitable per se (Reinartz and Kumar 2000; Libai et al. 2002).

6.2.3 Value of Initial Customer Cohort

Thus, the value of the initial cohort of customers is the sum of the values of the initial individuals. This implies that

$$CE_0 = \sum_{k=(v_{0-1}+1)}^{v_0} \sum_{t=0}^{T} r_i^t \frac{(R_{ti} - C_{ti})}{(1+d)^t}. \tag{6.2}$$

6.2.4 Total Customer Equity

Accordingly, the total customer equity is interpreted as the present value of all T cohorts, which yields

$$CE = \sum_{s=0}^{T} \frac{1}{(1+d)^s} \sum_{k=(v_{s-1}+1)}^{v_s} \sum_{t=s}^{T} r_i^t \frac{(R_{ti} - C_{ti})}{(1+d)^t}, \tag{6.3}$$

with the customer equity CE, the index s over customer cohorts, the forecasting periods T, the opportunity costs of capital d, the size of a cohort k at the end of the period v_s, the retention rate r_i, the revenues R_{ti} per customer i in period t, and the costs C_{ti}. While the last sum captures the lifetime value of an individual customer, the second sum is equal to the lifetime value of all new consumers. The third sum is the present value of all cohorts and therefore equal to the total customer equity.

6.2.5 Customer-Based Corporate Value

Based the outlined considerations, the customer-based corporate value is defined as the present value of the cash flows deriving from the current and future transactions with all individual customers. This is equal to

$$CV = CE - \sum_{t=0}^{T} \frac{FC_t + InvWC_t + InvFC_t + Tax_t}{(1+d)^t} + \frac{CV_T}{(1+d)^T} + NA - D, \tag{6.4}$$

in which CV represents the corporate value, CE the customer equity, t the time index, d the cost of capital, FC_t the fixed costs, $InvWC_t$ the net investments in working capital, $InvFC_t$ the net investments in fixed capital, Tax_t the tax payments, CV_t the terminal value, NA the value of the non-operating assets, and D the market value of debt.

6.2.6 Discussion of the DCF CEV-Model

The model resolves some of the research problems in valuation of software markets, but it is also restricted. It considers the value of cross-selling potential, but also reveals some vital limitations as sales growth, profit margin and the duration of the value-creating growth are simplified. An additional problem is the derivation of the terminal value, which is frequently approximated by a constant growth rate or a perpetuity. While the assumption of an annual purchasing cycle may be reasonable for some software segments, such as tax consulting software, it is unrealistic for other segments, such as operating software and office application suites (Reichheld 1999; Doyle 2000). Furthermore, the model assumes discrete purchases at the end of a period and a constant customer retention rate although migration models illustrate that purchase propensities may change (Dwyer 1997). Further limitations result from the data requirements of the model, as it is necessary to determine the past and current number of customers, the gross margin, the growth rates, the marketing expenses, the risk premium, and the costs of capital (Gupta et al. 2001). While some data should be available, other information is likely to be restricted and has to be approximated. Although retention rates are critically dependant on the specific context, research frequently assumes that they vary between 80 and 90 percent for established firms (Reichheld 1999).[4] Once the margin is determined, its growth is frequently approximated by a constant growth factor (Gupta et al. 2001). The costs of capital are usually calculated based on standard capital market models such as the CAPM (Brealey and Myers 1996; Ross et al. 1996). This discussion reveals that the model is a contribution to the investigation of valuation in software markets. But at the same time it reveals that the properties and dynamics of customer networks, are not considered adequately. The customer network and the acquisition rate contain further insights into growth dynamics, as referrals influence the product diffusion process in software markets (Hogan et al. 2002). An additional limitation derives from the assumption of homogeneous cohorts of customers, as real-world customers are heterogeneous. Some of the outlined limitations are resolved in the real options version of the customer equity model.

6.3 Real Options Customer Equity Model

Alternatively to the outlined DCF version, it is also possible to develop a customer-centric valuation model based on a real options perspective. In both models the description of the underlying customer network is a vital issue. The inverse mean reversion model is particularly relevant to valuation in software markets, as it was designed originally for the valuation of growth companies (Krafft et al. 2002). In this model the development of the customer base is described by an *inverse mean*

[4] The number of *acquired customers* can be derived as the difference between the total number of customers at the end of a period and the defected customers or the Blattberg-Deighton model can be applied (Blattberg and Deighton 1996; Skiera 1999). The *acquisition margin* can be derived by dividing the total marketing costs by the number of new customers (Reinartz and Kumar 2000). The *contribution margin* can be approximated by dividing the EBITDA-margin by the total number of customers at the end of the year (Skiera 1999).

reversion process.[5] The model accounts for positive and negative network effects as a perfect correlation between customer base and net cash flows is assumed. This implies that because of the network effects successful companies become even stronger if they can rely on a strong customer base (Shapiro and Varian 1998). The consideration of positive and negative network effects is a vital advantage in comparison to the outlined DCF model.

6.3.1 Inverse Mean Reversion Model

Research reveals a specific growth profile of software companies that induces particular financing requirements (Shapiro and Varian 1998). Initially, the number of customers grows exponentially, but after a certain saturation the growth rate tends to decline and to converge towards an industry specific mean (Krafft et al. 2002). The inverted mean reversion approach models the growth rate as a stochastic variable with specific characteristics. While a mean reversion process returns to a mean value, the inverted mean reversion process assumes that the observed variable drifts away from a critical mean (Neftci 2000). The discrete stochastic representation of the customer development is assumed to depend on five parameters, such that

$$\triangle K_z = a \cdot [\bar{K} - K_z] \cdot \triangle + \sigma(t) \cdot z(t) \cdot \sqrt{\triangle t}, \tag{6.5}$$

where the change of the number of customers in state z is denoted with $\triangle K_z$, the length of the time step with Δt, the maximum and the minimum state with L and $-L$ respectively, while the speed of the divergence is a. It is assumed that $a < 0$ for $-L \leq z \leq L$ and $a = 0$ for $z > L$ or $z < -L$. (Krafft et al. 2002) This implies that the divergence increases with increasing divergence of the state variable from the critical value. The second term of the equation is stochastic. It consists of the state variable $z(t)$ and the volatility σ of the change in customers. If z exceeds $\mid L \mid$, the intensity parameter of the divergence a is set to 0, as no further changes are expected. In a binomial product diffusion model each state represents an upward or downward movement that increases or decreases the respective cash flows by the jump width k, while the number of customers increases or decreases by k. This implies that the number of customers K_z varies and K_0 is the initial number of customers. A sequence of upward and downward movements yields a recombining tree. The model accounts for network effects as the probabilities depend on the size of the customer base. Accordingly, an upward movement Π_z in state z is determined by

$$\text{for } z < L : \Pi_z \cdot (K_{z+1} - K_z) + (1 - \Pi_z) \cdot (K_{z-1} - K_z) = a \cdot (\bar{K} - K_z), \tag{6.6}$$

[5] In general a *mean reversion process* is a stochastic process which over time returns to its mean value.

where $\bar{K}$ is the critical mass, a is the speed factor and

$$\text{for } z > L : \Pi_z = 0.5. \tag{6.7}$$

Hence, the probability state can be described as

$$\Pi_z = \frac{a}{2k} \cdot (\bar{K} - K_0) + \frac{1 - a \cdot z}{2}. \tag{6.8}$$

The equation can be simplified to $\frac{1-a \cdot z}{2}$. It implies that the customer base grows if the installed base is large and shrinks if it is small. This property reflects the reinforcing and moderating dynamics of network effects in customer networks. The constant L is chosen such that the probability p_z is defined between 0 and 1.

6.3.2 Customer Equity Model

Based on the outlined considerations, the development of the customer base can be modeled by performing the outlined calculations for each node in every period. For this reason the overall value of a customer in the real options customer equity model is broken down into a direct and an indirect component in line with the previously outlined customer equity theory (Krafft et al. 2002).[6] In this customer equity model the annual cash flows increase by a growth rate g such that

$$C_z = C_0 \cdot (1 + g)^z, \tag{6.9}$$

where C_z are all cash flows generated by a customer in state z and C_0 is the cash flow generated by initial customers. The expected present value of the customer base is the sum of the cash flows of all customers discounted at a risk-adjusted discount rate. Accordingly, the value of all customers $V_z(t)$ in state z is the expected discounted value of all customers in the following period plus the cash flows of the current period C_z is

$$V_z(t) = \frac{\pi_z \cdot V_{z+1}(t+1) + (1 - \pi_z) \cdot V_{z-1}(t+1)}{1 + r} + C_z, \tag{6.10}$$

while the value in the last period T is

$$V_z(t) = C_z. \tag{6.11}$$

[6] *Direct economic returns* are the net present value of the customer during the total customer lifetime, while *indirect returns* reflect network effects of customer referrals.

The decisive parameters of the model are the volatility of customer fluctuations and the probability distribution of the customer network. As far as the probability distribution is concerned, the model is based on the assumption that the number of customers is either extremely high or low in comparison to the critical value. This implies a particular shape of the probability distribution in the form of a bathtub, in which extreme values can be observed more frequently than mean values (Krafft et al. 2002). This is a reasonable assumption for several segments of the software markets, as they either convince a critical mass of customers or vanish from the market. Simulations indicate that the interval within which the probability can be derived without regarding the state variable decreases with smaller a. Consequently, the bathtub probability distribution can be constructed by choosing the corresponding divergence speed a (Krafft et al. 2002). In the binomial model, each state is followed by two states. As each branch is represented by one equation, two requirements have to be considered. First, the pre-specified expected change in the number of customers has to be met and all probabilities have to sum up to one (Krafft et al. 2002). Some stochastic models have a third requirement with respect to the volatility of the underlying stochastic process, which would require a trinomial model (Hull and White 1994). The underlying bathtub distribution assumes a high degree of uncertainty close to the critical mass of customers and an even higher degree of volatility if K is far below or above the critical mass. The volatility process is highly state-dependent, as the volatility increases close to the critical mass. After a short period of time uncertainty is resolved whether the company is capable of reaching the critical mass of customer or not. This state dependent, but not time dependent, specification of the volatility is coherent with the characteristics of many software companies as the profile of the software companies reveals. The behavior is an indication of a phase or state-transition which would change the step width that could lead to a breakdown of network.

6.3.3 Numerical Example of Customer Value

The subsequent numerical example illustrates the difference of the outlined model to conventional valuation approaches (Krafft et al. 2002). First, a recombining tree is created for four periods based on the following assumptions. Initial annual revenues are assumed to be EUR 1, while the annual growth rate g is set to 5 percent and the initial cash flow per customer C_0 is assumed to be EUR 1. From a traditional DCF perspective, the present value of a simply assumed annual growth despite of opportunity costs is EUR 3,2. In contrast, the average value per customer according to the inverted mean reversion model is EUR $5,42$ based on the subsequent calculations (Tables 6.1–6.3).

The tables reveal the numerical values in EUR for the jump width k, growth rate g and speed a. A comparison of the results illustrates that the sum of direct and indirect customer value are substantially higher than the direct present value. As the model can also account for state specific events, such as bankruptcy, it is

Table 6.1 Sensitivity of numerical example to jump width k

Jump width (k)	Customer value (EUR)
20	4,31
40	4,51
60	4,92
80	5,42
100	5,51
200	8,41

Source: Author based on (Bauer and Hammerschmidt 2005)

Table 6.2 Sensitivity of numerical example to growth rate g

Growth rate (k)	Customer value (EUR)
5	5,42
10	5,85
15	6,33
20	6,84
30	7,98
60	12,58

Source: Author based on (Bauer and Hammerschmidt 2005)

Table 6.3 Sensitivity of numerical example to speed a

Speed (a)	Customer value (EUR)
0,0	4,81
−0,1	4,98
−0,2	5,19
−0,3	5,43
−0,5	5,59
−1,0	6,55

Source: Author based on (Bauer and Hammerschmidt 2005)

more flexible and more reliable than conventional models. A sensitivity analysis with respect to variations of the input variables reveals that it is possible to determine a correlation between the jump width k and the respective value of the customer (Krafft et al. 2002). This implies that the value of software companies augments with increasing risk, which is counterintuitive to conventional financial intuition. The explanation is that there is an asymmetric set of opportunities due to the optional character of the managerial flexibilities. While cash flows increase exponentially if the company is successful, shareholders can lose up to the invested capital. A complementary finding of the sensitivity analysis is that the value of the customer base increases with an increasing divergence of the speed factor a from the critical value $\bar{K}$. In financial terms, shareholders benefit from an asymmetric real option that increases in value with an increasing exposure to uncertainty (Trigeorgis 1996; Levett et al. 1999).

6.3.4 Discussion of the ROV CEV-Model

The investigations underline the necessity of an accurate description of the customer base and the respective customer network. In the outlined approach, growth is described as a stochastic process. Hence, the vital task is to identify a suitable stochastic process that adequately mimics the development of customer relationships in software markets. According to the outlined inverse mean reversion model, the growth rate follows a "bathtub distribution" that captures reinforcing and moderating impacts of network effects. As such phenomena are frequently observed in software markets, the model is a contribution to the investigations on customer network-centric valuation in software markets by resolving some of the identified limitations of conventional models. But the assumed probability distribution implies the assumption that the customer base develops either far above or below the critical value. In contrast, the previously outlined empirical research on network effects in software markets revealed that a company can also have a stagnating customer base close to the critical mass for a certain time. In addition, software companies rarely start their business close to the critical mass, and a considerable percentage of software companies never reaches it. This discussion reveals that a vital issue is the description of the growth rate, i.e. the development of the customer network with an appropriate stochastic process. But, in addition, a variety of further research questions remain, such as the determination of K, the calibration of k, and the analysis of the volatility of the number of clients. If the jump widths of up-state and down-state events is characterized by different jump widths per node, as in the case of a network breakdown, the tree loses the recombining property which make the calculations much more difficult. Hence, it is necessary to customize the model and the underlying stochastic process to the specific situation in the software market.

6.4 Reconsideration of Customer Equity Valuation

All in all, the outlined customer equity models are valuable contributions to the research question, as they account for network effects by considering the referral value of customers. At the same time they underline the importance of network effects and of the customer network for valuations in software markets. A comparison of the methodologies reveals that conventional valuation models are not capable of accounting for network effects and the differences can be explained by considering the managerial flexibilities as additional real options. Hence, the models underline the previously identified research gaps, but also reveals that further investigations are required. These are particularly necessary with respect to the development of the growth rate and the underlying product diffusion processes in software markets as they describe the vital value driver for valuation in software markets. Hence, further research on adoption and diffusion models is conducted in the following chapter.

Chapter 7
Adoption and Diffusion Models for Software Markets

> "Diffusion is a kind of social change, defined as the process by which alteration occurs in the structure and function of a social system. [...] most people depend mainly upon a subjective evaluation of an innovation that is conveyed to them from other individuals like themselves who have previously adopted the innovation."
>
> (Rogers 1995)

Previous research revealed that network effects represent flexibilities in customer networks of software companies which can be valued from a customer network-centric perspective by adjusting the growth rates of DCF models or by modeling them explicitly as real options. In both cases a core problem is to project the development of the customer network in order to determine the cash flow distribution of the software company. This problem is the main research focus for the rest of the book. Hence, relevant adoption and diffusion models are investigated in the following chapter. The goal is to identify suitable concepts that allow to model developments of customer networks based on which the respective cash flows could be derived. In a first step, macro-level software diffusion models are outlined, before the focus is shifted to microlevel adoption models. A reconsideration of the models concludes the chapter.

7.1 Software Diffusion Models

The development of a customer network can be interpreted as a spread of products in software markets from a macro diffusion perspective.[1] Diffusion research has a long tradition in a variety of disciplines such as Mathematics, Geography, Physics, Biology and is increasingly applied in social sciences (Mansfield 1961; Granovetter 1978; Valente 1995). Innovation diffusion research has a more specific focus on the description of product diffusion processes that roots back to two empirical studies

[1] A *diffusion* is defined as a dispersal process of an information, a disease or a product by which an innovation is adopted (Rogers 1983). Please note the difference between adoption and diffusion processes. The diffusion process is the ex-post result of adoption decisions.

A. Kemper, *Valuation of Network Effects in Software Markets*,
Contributions to Management Science, DOI 10.1007/978-3-7908-2367-7_7,

in the 1950s (Valente 1995; Marwell et al. 1988; Jansen 1999). In the first study the social network of four cities was investigated in order to analyze the impact of social networks on medical advices and drug development (Coleman et al. 1957). This study revealed that the drug diffusion is much faster among doctors with an integrated social network. In a second study, Rogers and Beal (1958) investigated the impact of social networks on the innovation adoption decision of farmers and found a highly significant relationship (Rogers and Beal 1958). Both pioneering studies indicate that social networks can have a significant influence on the diffusion of innovations (Schmalen et al. 1993). Based on such observations, researchers developed a variety of analytical and numerical diffusion models, some of which are investigated in the following. In the first step, some existing basic analytical models are presented (Bass 1969; Mahajan et al. 1990; Rogers 1995). Then, an advanced Markov matrix diffusion model is developed which is based on an Eigenvalue analysis of Markov matrices before relevant adoption models are presented in the following section.

7.1.1 Fundamental Diffusion Models

The fundamental analytical economic diffusion models are based on the installed base paradigm, according to which the number of adopters is a function of the existing network size.[2] Hence, the diffusion is determined by vertical feedback loops such as increasing returns or a critical mass of customers.[3] More formally, the installed base in period t can be described by the differential equation

$$N_t = g_t \cdot (M - N^*_{t-1}), \tag{7.1}$$

where N_t is the number of adopters in period t, M is the cumulative number of potential adopters, and N^*_{t-1} is the cumulative number of adopters until period $t-1$. Rearrangements for g_t at time t yield the diffusion coefficient describing the relationship between the rate of diffusion and the number of potential adopters such that

$$g_t = \frac{N_t}{M - N^*_{t-1}}, \tag{7.2}$$

which is determined by the nature of the innovation, the available communication channels, and the specific characteristics of the social system (Mahajan and Petersonf 1985). The coefficient can also be expressed as a function of the cumulative adopters, such that

[2] This is a suitable assumption for modeling software markets as it is in line with the outlined properties. Please confer Sect. 5.3.

[3] This phenomenon is also known as frequency dependency effect.

$$g_t = a + b \cdot N^*_{t-1}. \tag{7.3}$$

Based on this generic diffusion equation, three relevant versions of the model are differentiated with respect to their diffusion behavior:

- Exponential Diffusion Model
- Logistic Diffusion Model
- Semi-Logistic Diffusion Model

All versions are investigated in the following in a finer level of detail before a Markov matrix diffusion model is developed (Bass 1969; Mahajan et al. 1990; Rogers 1995).

7.1.1.1 Exponential Diffusion Model

This *exponential diffusion model* can be applied to model a rapid exponential diffusion process. It is also known as an *external influence model* since the number of new adopters is assumed to depend exclusively on external influences, e.g. mass communication (Weiber 1993).

$$N_t = a \cdot (M - N^*_{t-1}). \tag{7.4}$$

The exponential diffusion model has only a limited potential to explain diffusion processes in software markets. They are governed by interdependent consumer decisions and network effects which are not considered in this model.[4]

7.1.1.2 Logistic Diffusion Model

This version is also termed internal influence model (Lilien et al. 1992). It is based on the underlying assumption that new adopters are primarily attracted by the existing installed base of customers due to word-of-mouth referrals.

$$N_t = b \cdot N^*_{t-1} \cdot (M - N^*_{t-1}) \tag{7.5}$$

The model is relevant for the description of diffusion processes in software markets as previous investigations revealed the considerable impact of word-of-mouth referrals due to network effects.[5]

[4] Please confer Sect. 3.3.11.

[5] Please confer Sects. 3.3.11 and 5.3.

7.1.1.3 Semi-Logistic Diffusion Model

The semi-logistic version of the generic diffusion equation accounts for internal as well as external factors. It can be described as a combination of both, such that

$$N_t = a \cdot (M - N^*_{t-1}) + b \cdot N^*_{t-1} \cdot (M - N^*_{t-1}). \tag{7.6}$$

The most popular semi-logistic diffusion model is the Bass Diffusion Model, also known as a mixed-influence model (Bass 1969). It has been applied successfully to retail services, industrial technology, as well as to agricultural, educational, pharmaceutical, and consumer durable goods (Mahajan et al. 1990). According to this model, a diffusion is described as

$$N_t = N_{t-1} + p(m - N_{t-1}) + q\frac{N_{t-1}}{m}(m - N_{t-1}). \tag{7.7}$$

Consequently, the number of adopters N_t at time t is a function of

Market Potential m. The total number of people who will eventually use the product.
External Innovation Coefficient p. Representing the likelihood that a potential customer starts to use the product due to external factors, e.g. mass media advertisement.
Internal Imitation Coefficient q. Representing the likelihood that a potential customer starts to use the product due to word-of-mouth recommendations.

This model provides some useful insights into network effects in software markets as it allows one to quickly derive a first understanding of the diffusion dynamics, while it does not account for complex dynamics due to its inability to consider the critical mass of a network, diffusion patterns and competition (Mahajan et al. 1990). Despite this criticism, the Bass diffusion model is a reasonable approximation in case of low consumer interdependencies and a moderate product diffusion rate (Schoder 1995). Since software markets reveal a high relevance of network effects and large diffusion rates, other diffusion models are required.[6]

7.1.2 *Markov Matrix Diffusion Model*

The reconsideration of the basic diffusion models revealed limitations. In the following, an advanced analytical approach is developed in order to overcome some of the outlined restrictions by investigating the long-term dynamics of diffusion processes based on a Markov matrix eigenvalue analysis.

[6] Please confer the findings in Sect. 3.3.

7.1.2.1 Principles of a Markov Matrix Eigenvalue Analysis

The *Markov matrix eigenvalue analysis* is an algebraic concept that was originally designed to find solutions to large scale systems of equations and eigenvalue problems.[7] For this purpose, the stochastic behavior of Markov chains represented in Markov matrices is analyzed based on an interpretation of their eigenvalues and eigenvectors.[8] A Markov matrix is characterized by two key properties (Strang 2003; Muthsam 2006):

1. All probabilities of a Markov matrix are either zero or positive and are defined in a range between zero and one.
2. All probabilities of a Markov matrix add up to one, as the total population is assumed to be conserved.

More formally, a *Markov Matrix* or *row-stochastic matrix* is a real $n \times n$ matrix $A = [a_{ij}]$ such that

(i) $a_{ij} \geq 0$ for $1 \leq i, j \leq n$,

(ii) $\sum_{j=1}^{n} a_{ij} = 1$ for $1 \leq i \leq n$.

In other words, Markov matrices represent probability or state vectors that contain nonnegative numbers and which add up to one. In general, an Eigenvalue analysis is an algebraic concept that allows to study internal properties of systems.[9] In the given analysis they are applied in order to study the long-term behavior and stability of Markov matrices by interpreting their eigenvectors and eigenvalues that contain vital information on the mixing behavior of populations. While the eigenvalue analysis is successfully applied to a variety of mathematical and engineering problems, it is a new idea to apply it as an analytical tool for studying diffusion dynamics in software markets. Accordingly, a Markov matrix eigenvalue analysis applied to study product diffusion in software markets contains the following five steps.

1. *Initialization of the Markov Matrix Diffusion Model.* The investigated model of a software diffusion process is formalized in a Markov matrix with static Markov probabilities for each possible state.
2. *First Iterations of the Markov Matrix Diffusion Model.* The first iterations can be computed by matrix multiplication that allows one to investigate the initial dynamics of the diffusion process.
3. *Long-Term Behavior of the Markov Matrix Diffusion Model.* In order to gain a deeper understanding of the diffusion process, the eigenvectors and eigenvalues

[7] Please confer Beutelspacher (2003), Strang (2003), and Muthsam (2006).

[8] A *Markov chain* is a sequence of probability vectors with a stochastic matrix.

[9] The *Eigenvalue analysis* states that given a square matrix $A_{n \times n}$, there exists a set of n scalar values λ and n corresponding non-trivial vectors v such that $Av = \lambda v$. In this context λ is termed the *eigenvalue* of A while v is coined the corresponding *eigenvector* of A (Beutelspacher 2003; Muthsam 2006). In other words, eigenvalues of a matrix m are the values λ_i for which one can find nonzero vectors v_i such that $mv_i = \lambda_i v_i$, while eigenvectors are the respective vectors v_i. Please confer (Strang 2003) or (Beutelspacher 2003) for further details on eigenvectors and eigenvalues.

of the Markov matrix are calculated in order to investigate a possible steady state of the system. Based on the assumption that the size of the total population is conserved, the final distribution of the population is a multiple of the population and its eigenvectors.
4. *Finite Iterations of the Markov Matrix Diffusion Model.* If all eigenvalues and eigenvectors are known the diffusion process can be calculated after finite time steps by solving a system of respective equations.
5. *Interpretation of Results.* In the final step of the analysis, the findings of the analysis are interpreted in the context of the investigations, i.e. the product diffusions in software markets.

In order to illustrate its contribution to investigations on product diffusions in software markets and therefore also to valuation in software markets, the Markov matrix eigenvalue analysis is applied in the following case study.

7.1.2.2 Markov Representation of a Software Diffusion Process

It is assumed that the investigated software market consists of a population, that can switch between two possible states, potential customer (U_P) and actual customers (U_A). This implies that the population is separated into two groups. First, potential customers, who do not own the software but could purchase it in the future. Second, the customers who have already adapted the software. The following analysis investigates the development over time of both groups. For this reason, the process is segmented into rounds. In each round both groups can switch their state. While some potential customers purchase the software, others will stop using it. Hence, the diffusion of the software is simulated by modelling the purchasing decision based on probabilities that should be derived from reliable empirical market research. Thereby, the diffusion process in software markets can be analyzed with the Markov Matrix Diffusion Model.

1. Initialization of the Markov Software Diffusion Model

In the following case study, it is assumed that potential customers have a 90% chance of becoming a customer while there is a 10% chance that they will remain potential customers. In addition, the probability is 20% that customers stop using the software, while there is, in turn, a probability of 80% that customers will continue their consumption.[10] Formally, the outlined assumptions can be formalized in a

[10] Please note that the sum of both states adds up to one as the diffusion process of this example is described by a Markov matrix with static markov probabilities. The design of this model implies that each potential customer can decide to purchase the product, i.e. that the product is known to all potential customers. However, a key feature of innovative software is that it is unknown to large parts of the population. Therefore, it is necessary to account for the increasing popularity and other in the setup of the diffusion matrix.

state matrix of a software diffusion process such that

$$\begin{bmatrix} U_A \\ U_P \end{bmatrix} = \begin{bmatrix} 0.9 & 0.2 \\ 0.1 & 0.8 \end{bmatrix} \begin{bmatrix} U_A \\ U_P \end{bmatrix}, \tag{7.8}$$

with $t = K + 1$. The size of the initial population is set to 1,000 and no initial users are assumed to be zero. All agents are potential customers and the size of the population is conserved over time.[11] Hence, the initial state of the software market model at $t = 0$ is

$$\begin{bmatrix} U_A \\ U_P \end{bmatrix}_0 = \begin{bmatrix} 0 \\ 1,000 \end{bmatrix}, \tag{7.9}$$

The diffusion process, including the initial conditions, can be described as

$$\begin{bmatrix} U_A \\ U_P \end{bmatrix}_k = \begin{bmatrix} 0.9 & 0.2 \\ 0.1 & 0.8 \end{bmatrix} \begin{bmatrix} U_A \\ U_P \end{bmatrix}, \tag{7.10}$$

Based on the outlined assumptions, it is possible to conduct a Markov analysis in order to determine the results of the diffusion process after a number of iterations and its long-term behavior.

2. First Iterations of the Markov Software Diffusion Model

First, the software market development is determined after the first iteration of the software diffusion process. Formally, it is necessary to solve

$$\begin{bmatrix} U_A \\ U_P \end{bmatrix}_1 = \begin{bmatrix} 0.9 & 0.2 \\ 0.1 & 0.8 \end{bmatrix} \begin{bmatrix} 0 \\ 1,000 \end{bmatrix}, \tag{7.11}$$

which is equal to

$$\begin{bmatrix} U_A \\ U_P \end{bmatrix}_1 = \begin{bmatrix} 200 \\ 800 \end{bmatrix}. \tag{7.12}$$

This result implies that 200 people adopt the software during the first round, while 800 remain potential customers. In a subsequent round this result of the initial round is multiplied again with the Markov matrix describing the decision criteria of the population in the software diffusion process. It is likely that even more people will adopt the software, while some users of the software will in turn stop using the

[11] It is important to be aware of this simplification, which is particularly relevant for the subsequent comparison to numerical network diffusion approaches. Please confer Chap. 7.3.

software. Hence, it is reasonable to assume that the number of adopters will grow above 200, while the number of potential customers will drop below 800.

3. Long-Term Behavior of the Markov Software Diffusion Model

In order to gain a deeper understanding of the diffusion process, the eigenvectors and eigenvalues of the Markov matrix that determine the decision process are investigated. Eigenvalues are the roots of the *characteristic polynomial*, which is defined for an $n \times n$ matrix as

$$Det\,[m - \lambda \times \text{Identity Matrix}\,[n]] = 0. \tag{7.13}$$

The n eigenvalues of an $n \times n$ matrix can be investigated by solving an nth-degree polynomial equation (Strang 2003; Beutelspacher 2003; Muthsam 2006). Therefore, the two eigenvalues of the Markov matrix in this case study can be determined as follows. As all Markov matrices have at least one Eigenvalue equal to 1, it is possible to deduct $\lambda_1 = 1$.[12] The second eigenvalue can be computed by calculating either the trace or the determinant of the matrix (Strang 2003; Beutelspacher 2003; Muthsam 2006).[13] Hence, the second eigenvalue is $\lambda_2 = 0.7$, which is below one.

In the next step, the eigenvectors are calculated based on the relationship

$$\text{Det}\,[m - \lambda \times \text{Identity Matrix}\,[n]] = \begin{bmatrix} 0 \\ 0 \end{bmatrix}. \tag{7.14}$$

(Strang 2003; Beutelspacher 2003; Muthsam 2006) Hence, the eigenvector X_1 is

$$\begin{bmatrix} 0.9 & 0.2 \\ 0.1 & 0.8 \end{bmatrix} \cdot \lambda_1 = 1 \begin{bmatrix} -0.1 & 0.2 \\ 0.1 & -0.2 \end{bmatrix} X_1 = \begin{bmatrix} 0 \\ 0 \end{bmatrix} \tag{7.15}$$

which is equal to

$$\begin{bmatrix} 0.9 & 0.2 \\ 0.1 & 0.8 \end{bmatrix} \cdot \lambda_1 = 1 \begin{bmatrix} -0.1 & 0.2 \\ 0.1 & -0.2 \end{bmatrix} \begin{bmatrix} 2 \\ 1 \end{bmatrix} = \begin{bmatrix} 0 \\ 0 \end{bmatrix} \tag{7.16}$$

Accordingly, the eigenvector is equal to

$$X_1 = \begin{bmatrix} 2 \\ 1 \end{bmatrix},^{14} \tag{7.17}$$

---

[12] Prove: Property (ii) implies $AJ_n = J_n$, with $J_n = [1, \ldots, 1]^t$. Therefore, 1 is always an eingenvalue of a Markov matrix A. q.e.d.

[13] Please note that the *trace* is the sum of the diagonal entries minus one. In this case $\lambda_2 = 0.9 + 0.8 - 1 = 1.7 - 1 = 0.7$.

Once the eigenvalues and a first eigenvector are known, a possible steady state of the system can be investigated. Based on the outlined assumptions, the size of the total population is assumed to remain constant. Moreover, it is important to note that the final distribution of the population is a multiple of the population and its eigenvectors. Based on the outlined calculations, the eigenvector X_1 implies that $\frac{2}{3}$ of the final population adopt the software, while $\frac{1}{3}$ of the population rejects it. Hence, the Markov matrix eigenvalue analysis suggests that based on the underlying assumptions 666.67 people adopt the software in infinity, while 333.33 will reject it.

4. Finite Iterations of the Markov Software Diffusion Model

In order to determine the state of the software diffusion model after finite diffusion steps, it is necessary to determine the other eigenvector X_2 by

$$\lambda_2 = 0.7 \cdot \begin{bmatrix} 0.2 & 0.2 \\ 0.1 & 0.1 \end{bmatrix} \cdot X_1 = \begin{bmatrix} 0 \\ 0 \end{bmatrix} \tag{7.18}$$

which is equal to

$$\lambda_2 = 0.7 \cdot \begin{bmatrix} 0.2 & 0.2 \\ 0.1 & 0.1 \end{bmatrix} \cdot \begin{bmatrix} -1 \\ 1 \end{bmatrix} = \begin{bmatrix} 0 \\ 0 \end{bmatrix} \tag{7.19}$$

and allows one to derive

$$X_2 = \begin{bmatrix} -1 \\ 1 \end{bmatrix}.^{15} \tag{7.20}$$

This allows one to calculate the solution after a hundred time steps:

$$U_K = c_1 \cdot \lambda_1^K \cdot X_1 + c_2 \cdot \lambda_2^K \cdot X_2 \tag{7.21}$$

which is

$$U_K = c_1 \cdot 1^K \begin{bmatrix} 2 \\ 1 \end{bmatrix} + c_2 \cdot (0.7)^K \begin{bmatrix} -1 \\ 1 \end{bmatrix}. \tag{7.22}$$

A second equation can be derived from the initial conditions

$$U_0 = \begin{bmatrix} 0 \\ 1,000 \end{bmatrix} = c_1 \begin{bmatrix} 2 \\ 1 \end{bmatrix} + c_2 \begin{bmatrix} -1 \\ 1 \end{bmatrix} \tag{7.23}$$

[14] Please note that the derived eigenvector is positive.

[15] Please note that the derived eigenvector is also positive.

such that it is necessary to solve a system with two equations, two unknowns, two constants and independent eigenvectors:

$$U_0 = \begin{bmatrix} 0 \\ 1,000 \end{bmatrix} = \frac{1,000}{3} \begin{bmatrix} 2 \\ 1 \end{bmatrix} + \frac{2,000}{3} \begin{bmatrix} -1 \\ 1 \end{bmatrix} \tag{7.24}$$

Therefore, the system approaches after 100 iterations the term

$$U_{100} = \frac{1,000}{3} \begin{bmatrix} 2 \\ 1 \end{bmatrix}, \tag{7.25}$$

while the second part of the equation

$$\frac{2,000}{3} \begin{bmatrix} -1 \\ 1 \end{bmatrix} \tag{7.26}$$

disappears.

5. Interpretation of the Markov Software Diffusion Model

Based on the underlying assumptions, the investigations based on the Markov Software Diffusion Model suggest that 666.67 people will adopt the software in the long-term, while 333.33 people in the customer network will reject it.

7.1.2.3 Reconsideration of a Software Diffusion Process

The case study illustrates that the developed Markov matrix diffusion analysis is another suitable analytical tool to investigate system states after finite steps and the long-term behavior of product diffusions in software markets. It strengthens quickly the intuition with respect to the dynamics in software markets as it reveals the results of the diffusion process after a specific number of iterations and over an infinite time horizon. Hence, it is an analytical contribution to the research problem based on the underlying simplifying assumptions.

7.2 Software Adoption Models

Adoption models have a research focus on the microlevel of individual purchasing decisions. Relevant adoption models for modeling the development of customer networks are reviewed. Adoption is the purchase of an innovation from a micro consumer perspective. Aggregated adoption decisions yield in a macrolevel diffusion

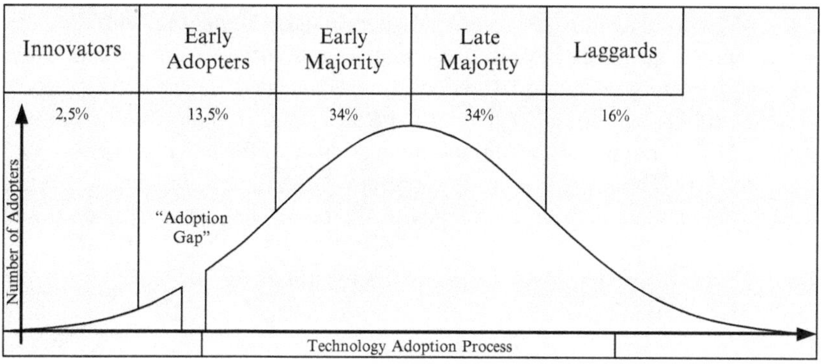

Fig. 7.1 Distribution Of Technology Adopters
Source: Rogers (1995)

process (Davis et al. 1989).[16] As this is a broad research area, only the most relevant concepts are depicted. First, the technology acceptance model is outlined that provides an overview of different classes of adopters, before centralized and decentralized standardization models are depicted.

7.2.1 Technology Acceptance Model

The Technology Acceptance Model (TAM) provides a generic overview on adoption by classifying various strata of adopters according to their attitude towards innovation. Accordingly, customers are clustered into five segments (Davis et al. 1989; Rogers 1995) (Fig. 7.1):

1. *Innovators*. This cluster of customers are pioneers. Usually, they are part of innovative cliques, have career security, control sufficient resources and have learned to cope with set backs. Such capabilities allow them to deal with even complex innovations despite high uncertainty. Innovators are the most important communication and distribution channel for innovations. According to empirical research studies they represent on average 2.5% of the total population.
2. *Early Adopters*. This group consists of opinion leaders willing to explore innovative ideas. Other people who follow their advice as they see the leaders as role models. Early adopters are capable of dealing with abstractions and frequently have a positive attitude towards change and uncertainty. Because of this attitude, and their influence, their choice can be decisive for the outcome of diffusion processes. Research indicates that this group represents on average 13.5% of the total population.

[16] Please confer Sect. 7.1.

3. *Early Majority*. The third group of customers consists of people, who are careful but accept change. They have frequent interactions with peers, but are rarely in a leadership position. Adoption decisions of this strata take longer as the customers prefer to follow advice rather than explore new venues. According to empirical studies 34% of the investigated population are part of the early majority.
4. *Late Majority*. The second majority represents skeptical people, who adopt innovations only if the majority are already using them. They want to be on the safe side. Hence, innovations are approached cautiously and their purchasing decisions are frequently based on peer pressure resulting from network effects. This group represents similarly to the early majority 34% of the total population.
5. *Laggards*. The final group consists of very conservative people. They are very suspicious and accept new ideas only if they are already mainstream as they want to be sure that an innovation does not fail in order to avoid stranding costs. This residual group accounts for an average of 16% of the total population.

All in all, the outlined model differentiates five types of customers based on their attitude towards innovation and uncertainty. It reveals the relevance of psychological acceptance parameters, such as perceived usefulness and perceived ease-of-use. Hence, purchasing decisions are constrained by a variety of factors, such as restrained cognitive abilities, time constraints, environmental and organisational limits (Ajzen and Fishbein 1980; Bagozzi et al. 1992).[17] The findings underline the importance of customer expectation management in software markets, which can be seen as one of the most significant success factors for adoption decisions along with compatibility, relative cost advantages, and complexity (Tornatzky and Klein 1982; Adams et al. 1992; Segars and Grover 1993; Subramanian 1994; Venkatesh and Davis 2000). Despite their importance, traditional valuation approaches do not typically account for such vital intangible factors which can have a decisive impact on the development of the customer network, the cash flows and thereby also on the respective value of the company. Customer network-centric valuation approaches should account for such properties by distinguishing various types of customers, their impact on the customer network, and on the respective cash flows.

7.2.2 Standardization Models

The adoption of software can be also be approached with standardization models (Miles and Snow 1978; Arthur 1989; David and Greenstein 1990; Schilling 1999; Hess 2000; Shy 2001). *Standardization* is an essential phenomenon in software markets that determines software diffusions based on cost-benefit considerations

[17] While the *perceived usefulness* is the assumed utility of a technology, the *perceived ease-of-use* represents the assumed relieve from effort.

of customers (Miles and Snow 1978; Buxmann and König 1998).[18] Accordingly, the cost-benefit considerations are investigated on a further level of detail. Standardized software facilitates the exchange of compatible data in customer networks. This increases the derivative utility of network effects. But it is also necessary to consider the related costs, e.g. higher switching costs and less competition. This cost-benefit consideration can be approached from a centralized or a decentralized perspective. While the centralized standardization problem assumes that a central unit coordinates the adoption decisions, the decentralized standardization problem is based on individual purchasing decisions (Buxmann 1996; Buxmann and König 1998; Buxmann et al. 1999).

7.2.2.1 Centralized Standardization Model

In the *centralized standardization model*, adoption costs and benefits of standardization are considered from the perspective of a central coordinating authority (Buxmann 1996; Buxmann and König 1998; Buxmann et al. 1999). In essence, the sum of the standardization and information costs are compared to the benefits resulting from a more efficient communication. More formally, G may denote the fixed generic utility of a software. If the derivative utility of network effects is termed D. The overall utility U is the sum of G and D.It is controlled by the binary standardization variable y_{ij}. This control variable is 1 if a pair of nodes ij shares a common standard or 0 if not. Hence, the overall utility of the centralized version is described as

$$\sum_{i=1}^{n}\sum_{j=1}^{n} G_i y_i + \sum_{i=1}^{n}\sum_{j=1}^{n} D_i y_i. \tag{7.27}$$

In turn, the adoption of a common standard is related to a fix price P_i. In order to model the costs, a binary indicative variable x is introduced, which is 1 if a program is used and 0 if not. In sum, the overall costs in centralized standardization models are equal to

$$\sum_{i=1}^{n} P_i x_i. \tag{7.28}$$

Comparing benefits and costs, the centralized standardization model assumes that software is purchased, if the sum of the utilities is at least equal to the respective costs. Hence, the adoption decision of the centralized standardization model can be expressed as

[18] In the following, *standardization* is defined as a set of rules and regulations governing the emergence and synchronization of common standards in software markets.

$$\sum_{i=1}^{n}\sum_{j=1}^{n} G_i y_i + \sum_{i=1}^{n}\sum_{j=1}^{n} D_i y_i - \sum_{i=1}^{n} P_i x_i \geq 0. \tag{7.29}$$

It is important to note that extensions of the model allow one to incorporate further parameters (Buxmann and König 1998). Nevertheless, it is not realistic to assume a central coordination unit with perfect information in software markets due to prohibitive information costs, heterogenous customers, incentive problems and the inability to reinforce centralized decisions in a free market economy. This implies that the centralized standardization model is a useful hypothetical construct that provides first-best standardization solutions given the underlying assumptions. But since individual purchasing decisions are more realistic for simulations of customer networks in software markets, a decentralized standardization model for adoption decisions in software markets is depicted in the following Sect. 7.2.2.2.

7.2.2.2 Decentralized Standardization Model

In the *decentralized standardization model* each node has the ability to make an individual adoption decision (Buxmann and König 1998; Buxmann et al. 1999). Due to the outlined startup problem, the costs for individuals frequently prevent an increase of users.[19] In order to determine the individual purchasing decisions and the respective critical mass, the decentralized standardization model is based on the following three critical assumptions:

1. *Individual Information.* The decentralized model is based on the assumption that network members know their costs, their individual utility, and the costs and benefits of their direct neighbors.
2. *Direct Connections.* It is assumed that the costs and benefits of directly connected network members can be estimated, or at least reasonably approximated, as they frequently exchange information.
3. *Indirect Connections.* As information about indirect contacts are frequently unavailable due to prohibitive costs, individuals assume that their costs and benefits are representative in order to approximate that of indirect connections.

Based on the outlined assumptions, all adoption decisions of the decentralized decision model are computed according to cost-benefit considerations and rational approximations. Hence, the expected derivative utility D of the network effects is equal to

$$E[D(i)] = \sum_{j=1}^{n} p_{ij} c_{ij} - C_i, \tag{7.30}$$

[19] Please confer Sect. 5.3.2.

with the probability p_{ij} that participant i assumes that another potential members adopt the software. A software is adopted if, and only if, the expected utility is positive or at least 0. Moreover, it is assumed that participant i has information about the costs c_{ij} if both are connected. Assuming that c_{ij} is representative for all other edges of j implies that the probability of connection p_{ij} is estimated by

$$p_{ij} = \frac{c_{ji}(n-1) - C_j}{c_{ji}(n-1)}. \tag{7.31}$$

The numerator represents the potential surplus of network effects of node j if all nodes adopt the software defined as utility minus costs, while the denominator normalizes the probability between 0 and 1. In other words, the probability that node j adopts the software product increases with an augmenting derivative utility relative to the price. In extreme, the adoption probability p_{ij} is equal to 1, if the adoption of the software is certain as node j only has advantages by the adoption. If, in turn, the numerator is negative as the costs are greater than the benefits, the adoption probability p_{ij} is zero. This decentralized standardization model provides a variety of features that are suitable for modeling adoption decisions in customer network-centric valuations in software markets. For this reason, a modified version of the decentralized standardization model is applied in the design of a customer network-centric valuation framework for software markets.

A comparison of the centralized and the decentralized standardization models reveals that it may be beneficial from a centralized perspective to implement a standardization, whereas individual purchasing parameters may prevent the adoption of a product. This can yield to inefficient market result and underlines the importance of the critical mass of customer networks. As consumers try to avoid such stranding costs, it can even be rational for competitors to agree on a common standard in order to jointly develop a market.[20]

7.3 Reconsideration of Software Adoption and Diffusion Models

After the description of the outlined adoption and diffusion models, it is necessary to assess their explanatory potential for customer network-centric valuation in software markets. Both approaches contribute to an increased understanding of customer networks as they are a good starting point for describing real world developments in software markets. The basic diffusion models are useful approximations, whereas the developed Markov matrix diffusion model is an analytical tool for a

[20] A recent example is the decision of Toshiba to abandon its HD-DVD standard in favor of the BlueRay standard. The goal of this strategic decision was to speed-up the development of the High Definition market despite of considerable sunk costs for Toshiba. Please confer (Welt 2008) for further information on the end of this war between various standards.

more complex description of customer networks. Similarly, the adoption models provide a variety of additional insights into the development of customer networks. The technology acceptance model provides a generic overview on different types of adopters. Adoption models, in turn, allow instigations of purchasing decisions on a finer level of detail. While the centralized version provides a theoretical optimal adoption behavior, the decentralized version assumes an individual perspective based on reasonable cost-benefit considerations. As individual adoption decisions are essential in the development of customer networks for valuations, a modified version of the decentralized standardization model is integrated in the following design of a customer network-centric valuation framework. Despite such contributions the models are constrained by several limitations. The outlined diffusion models provide rather very generic contributions as they are restricted by rigorous assumptions. Hence, the focus is instead on the adoption models. The outlined adoption models have to be adopted as they were designed to solve coordination problems in rather small networks. They are restricted by the following limitations:

1. *Complexity*. Complexity problems arise with an increasing size in the network. Analytical solutions can be derived for small networks, but even numerical simulations are limited as computational power has increased tremendously, yet remains limited. Previous research has applied similar concepts only for relatively small networks, and without a theoretical background in the network theoretical investigations, e.g. (Buxmann 1996), (von Westarp 2003) or (Weitzel 2004). Hence, one of the key challenges in developing quantitative models of software markets for respective valuations is to develop an efficient scalable model that is also capable of handling large-scale networks.
2. *Circularity*. Previous research in software markets reveals that network effects have an impact on purchasing decisions. But since this factor is unknown in advance, expectations concerning the market development are decisive decision parameters. It is assumed that everybody decides whether or not to purchase the software based on a comparison of the individual utility with the respective individual costs. Hence, each market participant is confronted with the problem of determining which product to use at which time. This is also coined the *genuine software purchase problem*. On the other hand, the purchase decision is also influenced by network effects. An increasing customer base increases the perceived utility derived by a software product. This effect can be particularly important for the purchasing decision if mouth-to-mouth referrals are essential distribution channels. Decisions in such diffusions of products are termed the *derivative software purchase problem*. Both effects together cause a circularity problem that underlines the importance of expectation management based on simulations that allow investigations of interdependencies.
3. *Data Sources*. It is difficult to obtain the required data, particularly for the quantification of product benefits. The data may be available for software that allows quantifiable cost savings, e.g. EDI or ERP software, but it is difficult to obtain for other software segments. The centralized model is based on the assumption that a central authority has access to costs and benefits of all network participant. This is a heroic assumption due to prohibitive costs for the required empirical

investigations. At the same time, such costs are a convincing reason to conduct simulations.

4. *Network Properties, Topologies, and Dynamics*. The outlined analytical adoption and diffusion models ignore the topology of networks by assuming implicitly fully connected networks. This assumption contradicts the empirical and theoretical findings in research on complex networks.[21] Similarly, vital properties and dynamics of networks are not considered. Hence, a decisive research question is if the consideration of such research contributions provides any additional insights for the outlined research question on valuation in software markets.

All in all, the outlined adoption and diffusion models are contributions to the research questions. Particularly, adoption models account for adoption decisions of individual customers based on cost-benefit considerations. But the outlined analysis also reveals limitations. In the original version, the models are neither capable of explaining the variety of products in software markets, nor of accounting for network properties such as heterogeneous preferences or varying network topologies. In the following, a network effects framework for valuations in software markets is designed in order to close this research gap. For this purpose the outlined decentralized adoption model is integrated into a valuation process framework. This combination allows to enhance the quality of valuation in software markets, due to a better understanding of the relevant network effects in the customer networks.

[21] Please confer Sect. 10.3 for details.

Chapter 8
Network Effects Valuation Framework For Software Markets

"Sponsor the access, charge the use!"
Principle of network management

The previous review of the valuation literature review revealed that currently there are no convincing solutions that account for network effects in software markets. The goal of the following chapter is to integrate the previously derived insights into a process framework for valuation in software markets. First, an overview of all phases of the framework is provided before the respective phases are depicted in detail.

8.1 The Network Effects Valuation Framework

The Network Effects Valuation Framework for Software Markets consists of the following four interdependent phases.

1. *Corporate and Software Market Analysis*. In the first step the corporation and the software market segment are investigated. The purpose of this first analysis is to develop a sound understanding of the business model based on which a software market model is developed in a second step.
2. *Software Market Model*. This market model can be used to investigate network effects and their impact on customer networks.
3. *Software Company Valuation Model*. Based on the insights from the outlined analysis the optimal valuation model is chosen and implemented.
4. *Sensitivity Analysis*. Finally, a sensitivity analysis that challenges the results concludes the framework.

It is important to note that it may be necessary to reiterate some steps of the framework as new insights in one phase may have an impact on the design of other phases. Particularly the final sensitivity analysis is intended to challenge the results of the valuation model and its setup. The interdependencies of the various phases are summarized in Fig. 8.1. In the following, each phase is depicted on a finer level of detail.

A. Kemper, *Valuation of Network Effects in Software Markets*, Contributions to Management Science, DOI 10.1007/978-3-7908-2367-7_8,

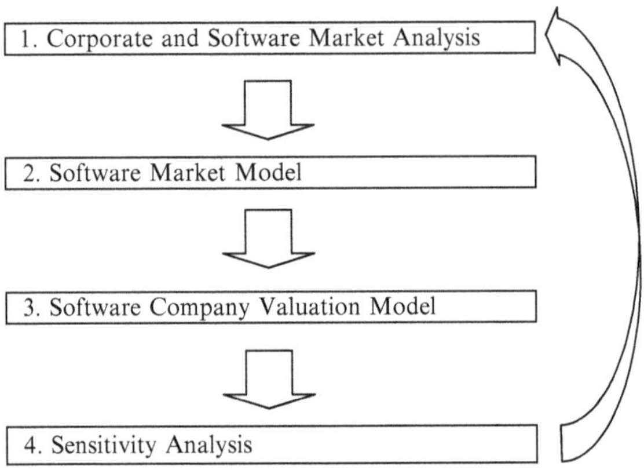

Fig. 8.1 Network effects valuation framework
Source: Author

8.2 Corporate and Software Market Analysis

The first phase of the framework is intended to provide an overview on the valuation target and its software market environment. For this purpose, the relevant characteristics of the market are taken into consideration before the company is analyzed. Here, the focus of the analysis is on the main value driver and on the cost structure of the company. Both investigations together allow one to determine the operating performance of the company from an investors' perspective.

8.2.1 Structure of Software Market

Depending on the underlying business model, different segments of software markets follow distinct rules.[1] In addition to the classical industry specific aspects of the structure-conduct-performance paradigm, it is also necessary to consider network effects due to their stated relevance for software markets.[2]

1. *Customers*. An analysis of the customers is central to all business models as they determine the demand for the product. Hence, it is important to understand the background, the market power and the cost-benefit considerations of customers. The previous analysis revealed that interdependent consumer decisions are frequently based on network effects.[3]

[1] Please confer Sect. 5.5.

[2] Please confer (Porter 2008) for the updated five forces industry model and for further information on the structure-conduct-performance paradigm.

[3] Please confer Sect. 3.3.

2. *Suppliers*. In classical industries the background, market structure and the incentive system of suppliers is important for an assessment of production restrictions. Software companies are rarely restricted by suppliers in the development, reproduction or distribution of software as these are frequently distributed via the Internet.[4]
3. *Competition*. According to the classical model, competition has a decisive impact on the supply and price strategy of companies. If the model is applied to software markets, it is important to consider the structure and the motivation of the direct and indirect competition as such aspects are likely to influence the competitive pressure as well. Actions and reactions of competitors may be integrated based on game-theoretical considerations, which in turn increase the complexity of investigations.
4. *Market Entry Barriers*. Market entry barriers represent potential changes for the competitive landscape due to the additional competition. This implies a game-theoretical uncertainty. If entry barriers are low and profits high, it is likely that other competitors will be attracted. In contrast, high market entry barriers such as large installed bases or high upfront investments, are likely to deter potential competitors.
5. *Substitutes*. Rivaling technologies which are perceived by customers as substitutes, are relevant as they are likely to influence the revenue streams of the software companies. If such substitutes are sufficiently attractive, customers switch the product at some point. Hence, substitutes have a decisive impact on those companies operating in software market segments whose are strongly driven by network effects.
6. *Network Effects*. As previously outlined, the underlying communication and interaction backbone, i.e. the customer network, has to be investigated in order to specify network effects. Possible are local interactions, i.e. word-of-mouth referrals, or implications from long-range interactions, i.e. media communication.

8.2.2 Value Driver of Software Companies

After this external market view the focus of the analysis is shifted to an internal company perspective. Based on a solid understanding of the software market segment, the main value driver of the company are considered. Depending on its maturity, the following main value driver of software companies can be identified (Wirtz and Kam 2001; von Westarp 2003; Bassen and Popovic 2004; Maaß and Pietsch 2008).[5]

---

[4] Please confer sect. 3.3.8.

[5] Please note the similarity to the phases of the previously presented product life cycle in Sect. 3.3.1.

8.2.2.1 Value Driver in the Innovation Phase of Software Companies

Companies in the innovation phase are startups with innovative products in the beginning of the company life cycle approach. Their financial performance is characterized by low revenues and losses. As they try to reach the critical mass in order to let their customers benefit from increasing network effects, the focus is on the customer network, the market position and the partner network.

8.2.2.2 Value Driver in the Expansion Phase of Software Companies

Companies in the expansion phase generate revenues and grow rapidly. Hence, revenues of the software company become increasingly important. In summary, the main value driver of this phase are the customer network, the market position, the partner network and the turnover.

8.2.2.3 Value Driver in the Maturity Phase of Software Companies

Mature software companies are established market player with stable cash flows generating profits, but with decreasing growth opportunities. Therefore, their growth rates decrease in comparison to the initial phases and to growing competitive startups. In this phase the classical value driver of company valuation are relevant. Due to their importance for valuation in software markets, particularly during the innovation phase of software companies, the revenues of the company are investigated in detail. Key to understanding the revenues of software companies is the relationship between software prices and sales, i.e. the price-quantity relationship of the business model.[6] For this reason it is necessary to investigate the sales and the pricing of the software on a finer level of detail, e.g. in a commercial due diligence (Niederdrenk and Maack 2008).

8.2.2.4 Sales of the Software

The quantity of sales are influence by a variety of factors.[7] Previous research reveals that the following aspects are most relevant to software markets.

1. *Potential Market*. Sales are determined by the maximum size of the market. This determines further growth and market opportunities as there is a limit to growth.
2. *Number of Customers*. The number of actual customers determines the contemporary revenues. It is highly relevant as software companies need a critical

[6] Please note that revenues are a product of sales volume times price.

[7] Please confer (Meffert 2000) for a broader overview.

mass of customers for a sustainable business model that is successful in software markets.
3. *Word-Of-Mouth Marketing.* Research on network effects reveals that software is frequently bought due to direct recommendations of the social environment, e.g. friends, colleagues or family. Hence, customer communication by word-of-mouth marketing is a decisive factor effecting sales that has to be evaluated individually.
4. *Product Features.* The product features are decisive factors for the purchase of software in software markets. Although programs tend to provide more functions than needed for the average user, e.g. Microsoft Excel or Word, users frequently require a minimum set of core functionalities.

The selection of parameters is not exhaustive and specific to the respective software market segments. However, the outlined factors are most relevant to valuation in software market and should, therefore, be incorporated in the design of a network effects software market model.

8.2.2.5 Price of the Software

Similarly, the price-quantity relationship is determined by the price of the software. This, in turn, is controlled by supply and demand. While the supply of software is virtually not restricted due to its availability via the Internet, the demand for software can be determined based on the following two important factors in software markets:

1. *Generic Utility of the Software.* In general, the willingness-to-pay of a customer is related to the perceived usefulness of the product. Hence, the generic utility of the software is a driver of the price.
2. *Network Utility of the Software.* As network effects determine the benefit of software, they influence also the among that a customer is willing to pay for it. The influence increases with a growing customer network.

According to empirical studies on the pricing of software, this list is not exhaustive but explains a significant percentage of the pricing mechanisms based on hedonic pricing models.[8] Hence, these factors should be incorporated in a software market model.

8.2.3 Cost Structure Analysis

In addition to the analysis of the topline, i.e. the revenues, it is also necessary to investigate the cost structure in order to determine the operative performance of software companies. The following cost factors should be taken into consideration in a software market model.

[8] Please confer Sect. 5.3 and (Groehn 1997) for further details.

1. *IT Costs*. Research reveals that software companies are frequently subject to a combination of high fix costs and marginal variable costs (von Westarp 2003). This constellation reinforces the emergence of network effects. The costs result mainly from the setup of a technological infrastructure and for required software licences.
2. *Labor Costs*. Labor is a vital cost factor for software companies as esearch, product development, and testing are very labor-intensive. The software has to be specified, developed and documented. This is done by well-paid software engineers who are either employees or freelancer.
3. *Marketing Costs*. Depending of the software market segment, the marketing costs for software can be marginal or gigantic. The increasing number of software companies induced increasing marketing costs over the last years, particularly if a company is new and unknown. As the costs to win a customer are six times as high in comparison to the costs for retaining a customer, the customer retention rate should be as high as possible (Kalkota and Robinson 2001). The customer churn rate, in turn, should be as low as possible.
4. *General Costs and Administration*. Another considerable cost driver of software companies are the rents, overheads and other fixed costs, e.g. office equipment.
5. *Expansion Costs*. During the expansion phase the costs can increase rapidly. While *modernization costs* are required to update the product functionality due to a relatively short lifecycle, it is also necessary to consider the costs for *extensions* of the customer network. Particularly, the costs for an adequate IT infrastructure and for labor can increase exponentially. Moreover, it is necessary to consider maintenance and update costs which can account for approximately 70 percent of the total software costs during the total product development cycle (Berger 2008). In addition, the increasing customer base may require technical upgrades in order to account for the larger number of customers, e.g. due to the necessity to open a call center or to hire external agencies and consultants.

Based on the value drivers and the respective cost structure of the individual software company, a software market model is designed in the following section.

8.3 Software Market Model

In the second step of the network effects valuation framework, the derived information is used to design a software market model. The goal is to investigate the properties and dynamics of the customer network to enhance the understanding of the underlying business model, as this allows one, to determine the critical mass of customers, the expected diffusion of the product and the expected distribution of the cash flows. The design of the model this based on the following interdependent phases, which integrate the most relevant aspects:

1. Scale and Scope of the Software Market Model
2. Implementation of the Software Market Model

3. Simulation of the Software Market Development
4. Derivation of Data for Valuation in Software Markets

In the following subsections, the four phases and their interdependencies are investigated in finer detail.

8.3.1 Scale and Scope of the Software Market Model

As a model is a representation of the real world, its scale and scope have to be determined before the model is designed. Depending on the optimal scale of representation, the size of the network can vary from relatively small, e.g. some individuals, to very large, e.g. groups, departments, or even whole nations. The following parameters determine the optimal scale and scope of the software market adoption model.

Network Size. The size of the population is equivalent to the number of interacting agents. As the complexity of numerical software market models increases exponentially, i.e. over-proportionally, with an increasing size of the network it is important to be aware of the complexity trade-off. The larger the simulated network, the higher the level of realism as well as the computational requirements. In general, the marginal contribution of an additional level of detail decreases with increasing level of detail. Hence, it is possible to compare the advantages and costs of various levels of representation in order to determine the most reasonable scale for the investigated software company.

Connections between Customers. As research on social networks reveals that social interactions in a customer network depend on their relationship to each other, the number and type of links between customers determines the level of detail of the software market model. The more complex the relationship, the higher the computational requirements. Prior research on social networks reveals that bidirectional links representing social relationships are a reasonable concept to model interactions among customers.

Benefits of the Software. According to the decentralized standardization model the incentive to purchase a software is the sum of a generic and of a derivative utility. While the generic utility is derived from the product itself, the additional derivative utility accounts for the identified network effects. Depending on the adoption rule, the description of the utility can decisively influence the scale of the model.

Costs of the Software. It is also necessary to quantify the opportunity costs of purchasing the software. Such costs result from the initial purchase or from additional maintenance or update costs. Similar to the modeling of the benefits, the level of detail of the adoption rule and the respective opportunity costs determine the scale of the model.

Once the relevant parameters are considered, the optimal scale and scope of the analysis can be determined. Other scales, e.g. larger or lower scales of representation, should be considered in a sensitivity analysis.

8.3.2 Implementation of the Software Market Model

If the scale and scope of the model is defined, it has to be implemented. This can be achieved by a numerical software market model with the following structure:[9]

1. Generation of a Customer Network Instance
2. Selection of a Software Adoption Algorithm
3. Storage of the Simulation Data
4. Analysis of the Network Data
5. Representation of the Network and Results

The five phases of implementation are described in detail in the following subsections.

8.3.2.1 Generation of a Customer Network Instance

In the first step of the implementation phase, it is necessary to generate a representative network instance, i.e. a network representation, in which the product adoption and diffusion processes are simulated. For this purpose it is assumed that each product has an installed base consisting of n independent actors within a population of p network participants who are connected by communication ties, representing friendships, family, or business relationships. The goal of this phase is to create a network instance with representative network properties.[10]

8.3.2.2 Selection of a Software Adoption Algorithm

In the second step of the implementation phase, the adoption decision of customers is codified based on the identified relevant decision parameters. This selection of a representative adoption algorithm is a crucial issue for the model which determines the reliability and the quality of the model. As there are a variety of approaches to model adoption decisions, it is necessary to select the most suitable approach based on the outlined relevant decision variables. Previous investigations revealed that a modified version of the numerical decentralized standardization model is a suitable

[9] Please confer also Chap. 11 and Sect. 13.3. This is a conceptual overview that is refined in the upcoming chapters of this book.

[10] Please note that the research on Complex Networks distinguishes a variety of network typologies which are discussed for didactic reasons in the next part of the book. Please confer IV.

model, as it accounts for network effects from a customer perspective.[11] Based on this decision the following refinements of the adoption rule have to be specified.

Set-Up Costs. Set-up costs are the initial purchasing costs. If competing products are considered, a switch to other software products may require the payment of set-up costs more than once. It is important to note that initial costs are sunk costs in following phases.

Maintenance Costs and Subscription Fees. In addition to the initial set-up costs, some software products are subject to additional subscription or maintenance fees.

Autarky Utility. Previous research revealed that a software can provide an autarky benefit. This is independent of other users as it describes the generic benefit of the software, e.g. a tax calculation program or a vocabulary training program.

Direct Network Effects. Direct network benefits result from an expanding customer network. This allows one to benefit from each neighbor directly by increasing possibilities for applying the software to network members. In turn, direct network effects are frequently also related to additional costs per link, e.g. time to add or maintain a link.

Indirect Network Effects. Indirect network effects occur also from an expansion of the customer network, but they are generated from indirect, i.e. not directly connected people of the customer network. They can also be related to additional indirect costs.

Net Benefit. The net benefit is the residual of the positive sum of the benefits deriving from autarky, direct network and indirect network effects minus the respective costs. Based on such decision parameters a net benefit coefficient can be computed. Please note that this cannot be negative as the software would not be purchased.

Estimation of Adoption Probability. Each agent has to form an opinion concerning the adoption probability of other users. There are a variety of approaches to resolve this game-theoretical problem. [12]

Duration of Adoption Process. The duration of the adoption process is a factor of primary interest. While in some software markets an equilibrium is reached after few iterations, others may require many iterations. Some will never reach a stable equilibrium. If the model reaches a steady state it is important to investigate the correlation of the network effects and the time until the steady state of the model. In this context it is possible to analyze the shortest diffusion path, which is the minimum amount of time that is required until a stable state equilibrium emerges in a given customer network. In other words, this can be the minimum time period that the company has to finance its operations in order to reach a critical mass of customers in the special case that the equilibrium is the critical mass and the net present value is positive.

---

[11] Please consider Sect. 7.2.

[12] Please confer Chap. 7 for a discussion of various adoption approaches.

Based on the outlined assumptions, the software market model allows one to approximate a distribution of revenues which is required in order to derive a reasonable distribution of cash flows for the software company. This, in turn, determines the valuations in software markets.

8.3.2.3 Storage of the Simulation Data

In a third implementation step, the results of the simulation have to be stored for subsequent analyses. Once the software market model is developed, it is iterated in order to simulate a variety of possible developments. There are several options to export the generated data, but since the data volume is significant in large customer networks it is important to implement efficient solutions. This comprises also an efficient interface for storing the data in a database for further investigations.

8.3.2.4 Analysis of the Network Data

The fourth step in the implementation is the analysis of the generated network data. This core element should provide a network theoretical toolkit for a static and dynamic network analysis. Hence, network theoretical tools are applied in order to derive network characteristics and to identify network dynamics of the analyzed software segment. Depending on the desired level of analysis and the required operations, the analysis provides more or less complex functionalities. Central is the number of adopters after a certain number of iterations.

8.3.2.5 Representation of the Network and Results

Finally, the network and the generated data are reported to the user. Depending on the size of the network, a variety of visualization options are available. In general, the goal is to provide in addition to numbers also visualization routines that depict the network, the adoption of the software and the findings of the network analysis, e.g. with Bayesian networks. Such a visualization of the software adoption and diffusion process allows an intuitive representation of the results and is also a plausibility check.

8.3.3 Simulation of the Software Market Development

Based on the designed network market model, multiple simulations should be performed in order to investigate the probability distribution of various outcomes. Each run represents a possible market development. Many paths are simulated in order to derive a representative probability distribution for the respective parameters. The results of the multiple runs are aggregated and interpreted with respect to their frequency. Such a procedure allows one to determine the probability distribution for various results. A variety of additional aspects can be relevant to the research question.

Individual Adoption and Diffusion Paths. Single adoption and diffusion paths can be depicted in detail. This allows the identification of the relationship of different network effects on each other. For example, it is possible to derive valuable insights into the process, such that the initial phase of a software adoption and diffusion process is instead determined by the autarky utility, while network effects tend to become more important in a later phase if the number of customers increases.

Relationship of Network Effects. As network effects can result from various sources, the relationship and proportion of all network effects is another interesting research focus. Accordingly, it is possible to outline the development of the indirect network effects in comparison to the direct network effects. Such insights, their relative strengths, and their development can be particularly relevant to valuations as they reveal insights on this potential value driver.

Total Net Benefit per Capita. Depending on the assumptions concerning the expected direct and indirect network utility, the total net benefit of the product can be computed per individual.

Once the simulations are conducted, the relevant data is derived for the valuation.

8.3.4 Derivation of Data for Valuations in Software Markets

In a final step of the implementation, the relevant data is collected for valuation purposes. Depending on the quality of the network data, the software market model allows to enhance the quality of the valuation. In this context the following parameters are particularly relevant to valuations in software markets:

Number of Adopters. The number of adopters allows one to describe a range of software users over time. This is particularly important if the goal of the analysis is to derive an approximation of the financial structure and performance at a specific point in time, e.g. after one year. Multiple simulations and sensitivity analyses allow to derive insights into the customer network size for various simulation parameter configurations.

Revenues. Based on the approximated number of customers, it is possible to derive a range for the respective turnover of the software company, if the prices of the software can be approximated.

Cash Flows. If relevant costs and depreciations can be approximated for the projection period, the outlined approach based on the software market model allows one to determine a range for the expected cash flows.

All in all, the insights derived from the outlined model contribute to investigate network characteristics and dynamics which are relevant for the subsequent valuation model. It is important to note that the outlined approach allows one to quantify the frequently neglected network effects. The additional information from this quantification can be useful for increasing the quality of the valuation, as depicted in the following section.

8.4 Software Company Valuation Model

In the third step of the valuation framework, the software company is valued based on the data generated from the software market model. In essence, the total value of the software company is split into a passive stand-alone value and an additional real option value. The passive value is the net present value, which can be frequently derived by a DCF approach. The additional optional value, in contrast, represents the value of managerial flexibilities such as network effects in customer networks. Such flexibilities are valued as options based on their underlying cash flow distributions, which can be approximated by the software market model. In essence, the software company valuation model is at this point a modified version of the classical real option valuation process that consists of four interdependent phases:[13]

1. *Identification of Main Sources of Uncertainty*. In the first step, the real option approach is framed. For this purpose asymmetric payoff structures are identified and indispensable characteristics of real options, such as flexibility, uncertainty and irreversibility, are investigated. Then, the scale and scope of the valuation model is determined based on the most significant sources of uncertainty.
2. *Selection of Option Pricing Model*. Once all relevant real options are identified, the most suitable valuation approach, a simulation, is identified and the corresponding option pricing model is selected.
3. *Determination of the Valuation Parameters*. In the next step the respective parameters of the selected option pricing model are determined with the simulation.
4. *Calculation of Option Values*. In a last step, the values of the options are determined and their interactions are considered. As the marginal contribution of additional options decreases, it is reasonable to concentrate on the most relevant options.

The outlined phases of the software company valuation model are summarized in Fig. 8.2 and described in finer detail in the following subsections:

[13] Please confer Sect. 2.3 for further details.

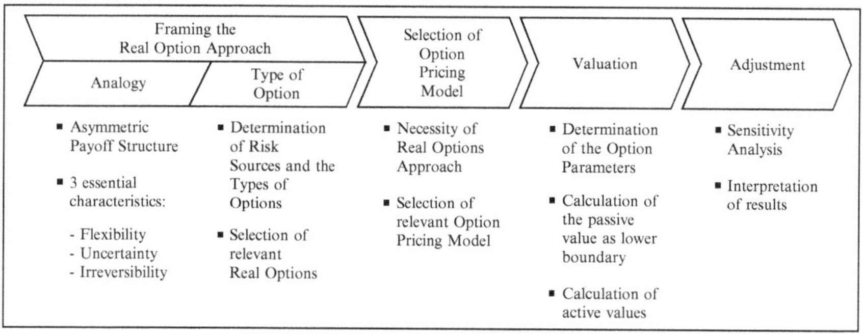

Fig. 8.2 Network effects real options valuation process. Source: Author based on Hommel (1999)

8.4.1 Identification of Main Sources of Uncertainty

In the initial phase of the software company valuation model, the main managerial flexibilities of software companies, such as customer networks, are identified based on a cost-benefit consideration and framed as options. For this purpose the respective volatilities are approximated with respect to the uncertainty of its underlying. An analysis of various software companies reveals that the following three types of real options appear most frequently in software markets:[14]

1. *Deferral Option.* Deferral options are options on cash flows in exchange for the initial investment providing a flexibility to extend the deadline before it expires. Investments in software companies frequently provide the flexibility to defer a management decision, e.g. the decision for a software standard, at the cost of an option premium.
2. *Scaling Option.* Scaling options provide the owner with the right to extend or to contract the production if market conditions require an adjustment of the production volume. In software markets the flexibility to extend the business in case of favorable market conditions can be decisive due to the outlined importance of network effects. An example is the purchase of a software company in order to leverage existing customer networks, e.g. for cross-selling or in order to reach a critical mass of customers.
3. *Liquidation Option.* Liquidation options are put options that allow the termination of an investment earlier than initially expected. In software markets the option to abandon a software company can be very valuable as it allows one to terminate unprofitable companies as soon as additional information on uncertain events becomes available, e.g. the manifestation of a standard.

[14] Please refer to Sect. 2.3.3 for the real options typology.

8.4.2 Selection of Option Pricing Model

Despite of the software company valuation model, the relevant option pricing model is chosen for the respective real option. Previous discussions revealed that numerical simulations based on the decentralized standardization approach are suitable for most valuations in software markets as they allow one to value even complex path-dependent payoff structures, which result from the individual software adoption perspective.[15] Hence, the outlined software market approach will be optimal for most valuations in software markets.

8.4.3 Determination of Valuation Parameters

In the third phase of the software company valuation model, the parameters of the selected option valuation approach are specified with respect to the investigated software company. As previously outlined, the value of options is determined by six parameters.[16] While some of them can be easily derived, others are more difficult to obtain (Hommel and Pritsch 1999):

1. *Price of the Underlying* V. The price of the underlying is the gross present value of the expected cash flows in the software market. It is derived by discounting the expected cash flows of the contractual flexibility at a risk-adjusted discount factor. Increasing present values have a positive impact on the values of options.
2. *Exercise Price* X. The exercise price of the option is the present value of the required initial investment costs once the option is exercised. It reduces the value of the option. The exercise price represents the required costs of the software company to exercise the option.
3. *Time to Maturity* t. This parameter is determined by the end of the investigated flexibility. In software markets the beginning and the end of a real option are frequently not contractually fixed and have to be approximated, e.g. by exogenous deadlines such as expirations of patents. If competition is involved, game-theoretic implications can be incorporated by adjusting the dividend parameter. In general, a longer time to maturity implies a higher option value.
4. *Volatility* σ. The volatility of the option is a measure of the uncertainty related to investment. It is one of the most important but also challenging parameters in the modeling of options. It can be approximated by distributions derived from the software market model. Contrary to intuition, increasing uncertainty raises the value of an option.
5. *Risk-free Rate* r. This parameter is a decision risk-free rate should have the same time to maturity as the investment object. Frequently, long-term government

[15] Please confer Sect. 4.3.

[16] Please confer Sect. 4.4.

bonds of low-risk countries are used as proxies for the risk-free rate. Increasing opportunity costs represented by the risk-free rate imply a decreasing option value.

6. *Dividends* D. Dividends are cash outflows representing leakages in the value of the underlying, such as increasing competition in software markets. Although it is difficult to incorporate dividends into option pricing models, they are applied if an endogenous game-theoretical modeling of competition is required. The dividends lower the value of an option and represent the opportunity costs if the option is not exercised.

Once all required parameters are derived or approximated, the value of the options can be computed.

8.4.4 Calculation of Option Values

In the fourth step of the software company valuation model, the value of the real option is computed based on the previously derived information. If all option pricing parameters are available, the value of the option can be determined by modified versions of existing option valuation approaches, such as the Schwartz and Moon model (Schwartz and Moon 2000). The Schwartz and Moon model were primarily designed to value Internet companies as a portfolio of options. It is based on a standard stochastic process in the form of a Wiener process. This describes its uncertainty, but the underlying mechanics to determine the required option valuation parameters can be applied to the framework for valuation in software markets. Nevertheless, it is important to recognize that a vital challenge in the valuation of software companies with real options is the derivation of the valuation parameters. Primarily the price of the underlying and the volatility. In the Schwartz and Moon model the volatility of the option is approximated as the expected cash flow distribution of stochastic simulations. And the respective expected cash flows are discounted at the appropriate discount factor in order to determine the value of the underlying. This transfers the problem to the identification of a suitable stochastic process that describes the development of the underlying. In this context the Schwartz and Moon model is an important research contribution that should be improved in order to reach a higher reliability. Hence, the fundamental idea is adapted to software markets and enhanced for the consideration of network effects. For this purpose the developed software market model based on adoption and diffusion models is used in order to understand the market dynamics, including the network effects, but also to find a suitable description for the underlying and its volatility. Due to the importance of network effects, this customer network-centric valuation in software markets is superior to a purely standard stochastic processes, e.g. the Wiener process or the inverse mean reversion process, as the software market model approach based on adoption and diffusion models accounts for network effects. Thus, the results of the previously conducted software market model can be integrated at this stage of the software company valuation model in order to compute an enhanced customer network-centric option value of software companies.

8.5 Sensitivity Analysis

In the final step of the network effects valuation framework, a sensitivity analysis is conducted. The goal is to challenge the findings of the software market model and the software company valuation model. Hence, the impact of one or multiple parameters on the results of the model are investigated in order to explore its stability and the most relevant value drivers. In addition, a profile of the software market is revealed that the segments have individual characteristics which have to be taken into consideration in the valuation of software companies. Hence, the results of the sensitivity analysis are cross-checked with respect to the characteristics of the individual software market segment. Further important issues of a sensitivity analysis are the interactions among multiple real options. These interactions can be first-order or second-order interdependencies (Trigeorgis 1996; Hommel and Pritsch 1999). First, it is necessary to consider first-order interactions, i.e. subsequent real options have an impact on the value of earlier options.[17] Moreover, second-order interactions have to be considered, i.e. the probability of an exercise of a real option has an impact on the value of the later real options.[18] After this depiction of the network effects framework for valuations in software markets, its advantages and limitations are reconsidered in the subsequent chapter. The findings of this discussion are the basis for further extensions and improvements.

[17] Please note that a simple sum of the option values would not be accurate, since the pay-off profiles of the respective underlyings would change simultaneously. Instead it is necessary to consider a synthesized underlying. Please confer (Trigeorgis 1996) for further details on the derivation of this synthesized underlying.

[18] The interactions refer to the change in the value of the underlying in dependence of the exercise of a real option. Hence, it may be necessary to account for the probabilities of exercise of previous options. This probability can be approximated by a partial differential equation with respect to the underlying of the closed-form solution in the form of a so-called delta. This states the probability of exercise for an option at the current moment. However, it is important to note that this delta-hedging is only possible for a specific set of parameters. Please confer Arditti (1996) and Hamilton (2000).

Chapter 9
Reconsideration of Network Effects in Software Markets

> "A crucial concept in understanding the nature of the diffusion process is the critical mass, which occurs at the point at which enough individuals have adopted an innovation that the innovation's further rate of adoption becomes self sustaining."
>
> (Rogers 1996)

The first important step was to identify the relevance of networks effects for valuations in software markets, as they are frequently ignored in the traditional literature. The failure to consider network effects can incite suboptimal underinvestment strategies, particularly if software companies operate in customer networks close to the critical mass. In order to develop a network effects framework for valuations in software markets, research on network economics, customer-equity valuation as well as on adoption and diffusion research was reviewed. The result is a framework that allows one to quantify network effects with software market models based on underlying adoption and diffusion models. In summary, the developed network effects model for valuation in software markets is the first step to close some of the identified research gaps, but it also reveals a variety of further research opportunities.

Heterogeneity of Agents. Although consumers have heterogeneous preferences, traditional research in network economics assumes typically homogeneous agents. But the heterogeneity of consumers is becoming increasingly important in order to understand the adoption behavior of customers (Valente and Davis 1999). For example, opinion leaders of customer networks are a phenomenon that cannot be explained by conventional network economic models.

Network Size and Complexity. Although a model is supposed to be a simplification of reality, it is also important to investigate the scaling of networks. Some researchers hypothesize that the size of a network is irrelevant in software market models, but do not support this hypothesis with convincing experimental or theoretical research (von Westarp 2003). Hence, research is required that investigates the scaling properties of the models.

Modeling Costs. The suggested network effects framework may require some additional efforts, but also reveals valuable insights that increase the quality and reliability of the model. Hence, its applicability has to be assessed on a case-by-case basis by investigating additional benefits and costs. The network theoretical

A. Kemper, *Valuation of Network Effects in Software Markets*,
Contributions to Management Science, DOI 10.1007/978-3-7908-2367-7_9,

assessments can be particularly important in the case of irreversible investment decisions, such as turnaround financing decisions, in which the optional value of the customer network is significant or even decisive.

Network Topology and Network Dynamics. The network effects model assumes that the average connectivity of customer networks in software markets can be described by a fixed number. From a network theoretical perspective, this assumption implies that the network is assumed to be a fully connected network, which is a simplification causing considerable consequences. In fully connected networks, monopolization is a common market result that can't explain the coexistence of standards as other structural network characteristics, e.g. network density or connectivity. Moreover, the outlined research underlines the importance to understand the dynamics of networks in customer networks for valuations in software markets, but it is assumed that the backbone of the underlying network exists and does not change over time. Its emergence and its dynamic development are not considered, yet. Hence, the discussion of network properties, topologies and dynamics is expanded in the following part.

Nature of Adoption and Diffusion Process. A crucial issue in the network effects framework is the description of the growth rate development for the option pricing. Further network theoretical analyses reveal that, in contrast to the standard description as a purely stochastic process, the adoption and diffusion process in software markets frequently follow instead a hybrid stochastic-deterministic pattern. While initially the software adoption process is best described by a stochastic process, it follows deterministic rules at a later stage of the diffusion process.

Some of the limitations are investigated in finer detail in the following part on modeling customer networks from a Complex Networks perspective.

Heterogeneity of Agents. In order to account for heterogeneous agents with individual adoption rules, the network effects framework is extended with a numerical agent-based network simulation to a complex networks valuation framework.

Network Characteristics. Research on complex networks reveals that networks share some common properties that can be quantified. The most relevant network characteristics are investigated in the following part.

Network Topologies and Dynamics. Complex networks research indicates that there are a variety of network topologies. As research reveals that the implicitly assumed random degree distribution for customer networks in software markets is frequently inappropriate, more research on network topologies and their respective dynamics is required (Albert and Barabasi 2002; Newman 2003b).

Based on this overview, the discussion on network properties, topologies and dynamics is expanded in the following part from a complex networks perspective.

Part IV
Modeling Customer Networks from a Complex Networks Perspective

The quantitative description of adoption and diffusion processes in customer networks with the network effects framework has been identified as a crucial success factor for customer network-centric valuation in software markets. But further research is required as its reconsideration also revealed a variety of limitations related to properties and dynamics of networks. For this purpose a complec networks perspective is assumed. The complex networks theory is a promising research perspective that aims at understanding properties, dynamics and topologies of real world networks. Accordingly, this part of the book analyzes customer networks in software markets in light of this paradigm. In the first step, an overview of the background of complex networks research is provided. Relevant properties, topologies and dynamics of complex networks are reviewed. Based on the derived insights, research hypotheses are formulated concerning the most relevant aspects of customer network-centric valuation in software markets. These comprise the variation of network topologies, scaling properties and the nature of customer networks in software markets. The investigations of these hypotheses are the primary focus of the rest of the book. For this purpose, a complex networks adoption and diffusion simulator is designed and implemented. Then, the selected hypotheses are challenged with complex networks research and with simulations. Based on the derived insights, the previously designed network effects framework is extended to a complex networks framework for customer network-centric valuations in software markets. Finally, a reconsideration of the research concludes this part of the book.

Chapter 10
Complex Networks Theory

> "Reductionism was the driving force behind much of the twentieth century's research. To comprehend nature, it tells us, we first must decipher its components. [...] Divide and conquer; the devil is in the details. [...] Now we are close to knowing just about everything there is to know about the pieces. But we are as far as we have ever been from understanding nature as a whole. [...] The reason is simple: Rising reductionism, we run into the hard wall of complexity. We have learned that nature is not a well-designed puzzle with only one way to put it back together. [...] It [Nature] does so by exploiting the all-encompassing laws of self-organization, whose roots are still largely a mystery to us. Today we increasingly recognize that nothing happens in isolation. [...] We have come to see that we live in a small world, where everything is linked to everything else. We are witnessing a revolution in the making as scientists from all different disciplines discover that complexity has a strict architecture. We have come to grasp the importance of networks."
>
> (Barabasi 2002)

This chapter provides a summary of relevant insights into complex networks theory. It is the foundation for the development of network theoretical hypotheses and the respective network methodology. In the first step, a brief overview on the background of complex networks research is provided, before relevant structural and locations properties of networks are presented. The network measures are used to illustrate insights into the structure and on the dynamics of networks. At the end of this chapter, research hypotheses are developed concerning the open research questions on network network-centric valuation in software markets. They are challenged with the complex networks diffusion simulator that is developed in the following chapter.

10.1 Principles of Complex Networks Research

The emergence of order in natural systems is a constant source of inspiration for research as most complex systems in nature do not offer a high degree of order. Many of the ordered systems form *complex networks* whose nodes are the elements of the system and edges represent interactions among the nodes.[1] This is

[1] *Complex networks* are networks with nontrivial network characteristics that differ from those of simple networks such as lattice rings.

A. Kemper, *Valuation of Network Effects in Software Markets*,
Contributions to Management Science, DOI 10.1007/978-3-7908-2367-7_10,

the focus of the emerging body of research on complex networks which investigates the underlying generic properties and dynamics of interdisciplinary complex networks (Newman 2000; Hayes 2000a; Hayes 2000b; Strogatz 2001; Albert and Barabasi 2002; Dorogovtsev and Mendes 2002; Newman et al. 2006). Complex networks research is at the cusp of research in several fields of inquiry.[2] Examples are applied mathematics, computer science, engineering, theoretical physics and theoretical biology. Complex networks is an active area of empirical and theoretical studies inspired by statistical properties of large-scale networks such as electrical power grids, the Internet, the World-Wide Web, telephone networks or academic citation networks, and a broad array of other practical issues (Albert et al. 1999; Faloutsos et al. 1999; Watts 1999; Amaral et al. 2000; Broder et al. 2000; Jeong et al. 2000; Fell and Wagner 2000; Williams and Martinez 2000; Newman 2001a; Liljeros et al. 2001; Jeong et al. 2001; Montoya and Sole 2002). The basic idea of this approach is that networks display universal non-trivial topological features, with patterns of connection between their elements. These are neither purely regular nor purely random, e.g. similar degree distributions, clustering coefficients or community structures.[3]

The history of complex networks research dates back to the origin of *graph theory*, a traditional branch of discrete mathematics. In 1735, Euler derived a solution to the Königsberg bridge problem that was based on topological abstractions by describing relationships and interactions between discrete objects (Harary et al. 1965). But at the same time, Euler's research contained the important message that even small changes in the topology can have vital implications on the behavior of the total network. In other words, the construction and structure of networks provide insights to understand the complex world around us. Graph theory boomed after Euler published his research with influential contributions from Cauchy, Kirchhoff, Huillier, Cayley, and Pólya (Cauchy 1813; Kirchhoff 1847; Huillier 1861; Cayley 1875; Pólya 1937). Their findings inspired various other streams of research on networks.

Social network theory has a research focus on social relationships and their underlying network structure, e.g. prestige, prominence, structural balance, clusterability, cohesive subgroups, and affiliation networks (Wasserman and Faust 1994). The research on social networks dates back to the 1930s, and became more formal in the 1950s, when research discovered the importance of degree distributions and that of other network characteristics (Moreno 1934; Rapoport 1957; Wasserman and Faust 1994; Scott 2000). A popular example is the small-world experiment of Milgram, stating that the average distance in social networks between randomly chosen individuals and a target individual are six degrees of separation (Milgram 1967; Travers and Milgram 1969).

The introduction of probabilistic methods in graph theory gave rise to another branch of research, known as the *random graph theory*, which has also been a

[2] Accordingly, *complex customer networks* are customer networks investigated from a complex network research perspective.

[3] Relevant properties of complex networks are explained in the following Sect. 10.2. Please confer (Newman et al. 2006) for further information on complex networks research.

fruitful source of graph-theoretical results. Particularly important are the discoveries of Erdos and Renyi in 1959. They revealed that random graphs are statistically homogeneous and have a particular degree distribution following a Poisson distribution (Erdos and Renyi 1959).

There have been dramatic advances in the past few years, driven by several parallel developments that caused the emergence of *complex networks research* (Albert 2001).[4] First, the computerization of data in all fields allowed the generation of large databases on the topology of various complex networks. Although most of the databases were not created with the purpose of studying networks, researchers could explore them to uncover the underlying network topology. A second important factor behind the advance of complex networks is increasing computing power. Thereby, it is possible to explore questions that could not be addressed a few years ago. Last but not least, the slow but noticeable breakdown of boundaries between disciplines has allowed researchers to explore interdisciplinary databases, bringing them a step closer to the underlying generic properties of complex networks. Hence, more recent complex networks studies indicate that topologies of networks also comprise scale-free networks, which are characterized by a power-law degree distributions and small-world networks with short path lengths combined with a high clustering coefficient (Newman 2003b). Further details on topologies of networks are expanded later after a description of the defining structural and locational properties of networks.[5]

10.2 Relevant Properties of Complex Networks

In the following section some definitions related to the properties of nodes and networks are provided. These are necessary in order to understand the definitions of the most relevant properties and topologies of complex networks.[6] First, some fundamental characteristics of individual nodes have to be defined. While many network quantities and measures have been proposed and investigated, the following structural and locational statistics are most relevant to customer networks-centric valuations in software markets:

- Degree, Indegree, Outdegree
- Degree Distribution

[4] Please note the difference between the social network theory and complex networks research. While the social network theory analyzes the characteristics of individual nodes in small networks, complex networks research investigates statistical properties of networks in order to observe large-scale structural and locational properties, as well as the large-scale network dynamics.

[5] Please confer (Barabasi 2002) for a more general overview on the development of complex networks research and (Newman et al. 2006) for a more technical overview on complex networks that contains all classical papers of complex networks research.

[6] There is much more that could be said about properties of complex networks, but a complete survey of all the material is beyond the scope of this paper. Please confer Newman (2003b) and Newman et al. (2006) for further information.

- Network Connectivity
- Network Centrality and Structural Equivalence
- Network Connectivity
- Traversity, Clustering and Density
- Assortative Mixing Patterns
- Degree Correlations
- Giant Component

These properties of nodes and networks, in turn, are the basis for investigations of network topologies and dynamics. Next, some basic definitions are provided below before the outlined properties are derived.

10.2.1 Fundamentals of Complex Networks Theory

We assume a network with n actors, which we model as a directed *Graph* $G = (V, E)$ with a set of nodes $V = 1, \ldots, N$ corresponding to the set of actors and a set of links $E = 1, \ldots, N$, for $E \subseteq V \times V$, where $(i, j) \in E$ if and only if actor j is influenced by actor i (Batten et al. 1995; Barabasi 2002; Albert and Barabasi 2002; Newman 2003b).[7] In software markets nodes represent economic agents, i.e. customers, while a link between two agents $i, j \in V$ represents a direct interaction between them such that $i, j \in E$. It is possible to denote links between agents by a binary variable $a(i, j) \forall i, j \in V$ such that $a(i, j) = 1$ if $i, j \in E$, or $a(i, j) = 0$ if not. It is important to note that the way nodes and edges are drawn is, in most cases, irrelevant. All that matters in the visual representation are the structural relationships in the network, i.e. which pairs of nodes are connected and which are not.

As the term *nonlinearity* is vital in this book, a solid definition is required. In essence, a problem is nonlinear if its solution cannot be written as a linear combination of independent components (Strogatz 1994). More formally, nonlinear behavior of dynamic systems is frequently described by differential equations such as

$$\frac{dx}{dt} = v(x), \tag{10.1}$$

[7] This terminology is typically used by computer scientists, while mathematicians speak of graphs that are defined by a set of nodes connected through edges (Simon 1962). The etymological origin of the word roots back to the Indogerman word "net" meaning "being tied" (Duden 1989). *Complex networks* have non-trivial topological features, i.e. features that do not occur in simple networks. The complexity of a system is related to the amount of information that is necessary to describe its behavior. The most fundamental subunit of a network is a *vertex*, which is also termed *site* in physics, *node* in computer science, and *actor* or *agent* in sociology. In contrast, an *edge* is the connection between two nodes, which is also termed *bond* in physics, *link* in computer science, or *tie* in sociology. Links are coined *directed* if they have an orientation, whereas bidirectional links are termed *undirected*. A set of connected describes a *graph*. Depending on the context, expressions may be used interchangeably throughout the subsequent research.

with respect to time t, where $x(t) = (x_1(t), \ldots, x_n(t))$ is a vector of state variables, and $v(x) = (v_1(x), \ldots, v_n(x))$ is a vector of functions describing the dynamics of the investigated system (Strogatz 2001). In such a system, long-term predictions are impossible due to amplifications of small uncertainties or measurement errors. Feedback loops cause the network to behave nonlinearly, i.e. a small change in one or more parameters can fundamentally change the behavior of the total system due to emerging properties that arise on different scales of observation. It is very difficult to solve such equations analytically. But it is sometimes possible to approximate the solution with an n-dimensional state space model that has the axes $x_1, \ldots, x_n$.[8] The dynamic development of a system can be interpreted as the trajectory of state spaces $x(t)$ (Strogatz 2001). Many real life systems exhibit such nonlinear dynamical behavior if they depend on inputs to the system as well as on the contemporary state (Verdult and Verhaegen 2000). Based on these definitions, the previously outlined properties of networks are derived in the following sections.

10.2.2 Degree, Indegree, Outdegree

Each node in a network has a specific *degree* or *connectivity* k that corresponds to the number of links connected to that node.[9] *Indegrees* and *outdegrees* of nodes describe the amount of incoming and outgoing edges. Accordingly, we denote by $in(j) := (i, j)|i \in V \cap E$ the set of actors that influence actor j. Conversely, we define as $out(i) := (i, j)|j \in V \cap E$ the set of actors that are influenced by actor i.

10.2.3 Degree Distribution

The *degree* of a nodes is defined as the number of incident connections (Erdos and Renyi 1959). Accordingly, the *degree probability* p_k of a network is the probability that a node chosen uniformly at random has degree k. It represents the fraction of node with degree k (Price 1965; Broder et al. 2000). The ensemble of all degree probabilities of a network in a histogram is the *degree distribution* of a network. A network is said to be connected, dense, or integrated if it is has a high degree

[8] A *nonlinear state space model* is a mathematical representation of a nonlinear dynamical system that accounts for the state of the system, as well as its past and future development. In other words, the state space is a set of all states a system can be found in. If the state space and the transition paths between the states are known it is possible to follow the dynamics of the system and to treat its behavior not only deterministically, but also in a probabilistic manner, which allows one to calculate and to understand time dependent physical properties. These are frequently determined as weighted averages over the whole state space, e.g. according to the mean-field approximation.

[9] Please note the difference between the network connectivity, i.e. connectivity defined for the whole network, and the general connectivity, i.e. the connectivity of a node. Please confer Sect. 10.2.5.

distribution, whereas it is radial if it consists of only a few connections (Wasserman and Faust 1994). The degree distributions of many real networks follow a power-law, e.g. the degree distribution of the webpages follows a powerlaw over several orders of magnitude (Albert et al. 1999). This implies a heterogeneous topology of the network in which the various nodes have different characteristics. While the majority of the nodes in the network has a small degree, there is a small fraction of highly connected nodes. Degree distributions are important measures related to network topologies as each topology has a characteristic degree distribution. Hence, they can be interpreted as "fingerprints" of networks.

10.2.4 Network Centrality and Structural Equivalence

Network centrality describes the relative importance of a node within a graph through its relative distance to other nodes, measured by the shortest path (Scott 2000). The lower the shortest path to other nodes, the higher the network centrality. *Structural equivalence*, in turn, implies that two different locations in the network are equivalent if they have a similar structural importance within the network, i.e. similar relationships and connections to other network participants.

10.2.5 Network Connectivity

Network connectivity describes the degree to which nodes are connected in a network, i.e. the intensity of physical or logical connections. It can be quantified based on network metrics such as the relative density, the shortest path, or the diameter of the network as depicted in the following sections.

10.2.5.1 Relative Network Density

Relative network density is defined as the total number of existing links T divided by the maximum number of links. A network with N participants has a maximum of $\frac{n(n-1)}{2}$ connections. In order to determine the relative density of a network, it is necessary to distinguish the density of symmetric and asymmetric networks. Formally, the network density $D(S)$ of *symmetric networks* is measured by dividing the total number of links T by the amount of maximum possible links. As the exact definition is different for symmetric and asymmetric network, the respective network densities are defined as:

(a) *Symmetric Networks*. For n participants, the maximum possible number of links in a symmetrical network is $\frac{n(n-1)}{2}$ and the respective density of the network is defined as

$$D(S) = \frac{T}{\frac{n \cdot (n-1)}{2}} \tag{10.2}$$

(b) *Asymmetric Networks.* The maximum possible number of links in an asymmetrical network is $n(n-1)$ yielding the density

$$D(S) = \frac{T}{n \cdot (n-1)} \tag{10.3}$$

A highly connected network is also termed a dense or an integrated network, whereas a network with low density is coined sparse or radial.

10.2.5.2 Shortest Path and Characteristic Path Length

The *shortest path length* $d(i, j)$ is the shortest connection between two nodes i and j in a network. It is the minimum number of nodes that must be traversed in order to reach node j from node i. This measure is also termed the *geodesic path* between two nodes. Please note that there is frequently more than one geodesic path between two nodes. Based on this definition the *characteristic path length* of a network is defined as the median of the shortest path lengths connecting each node to all other nodes (Watts 1999).

10.2.5.3 Network Diameter

Based on the definition of the geodesic, the *diameter* of a network is equal to length of the longest geodesic path between any two nodes of the network.[10] It is measured as the number of required edges. Frequently, the lowest and the highest distances between all nodes in a network are particularly relevant.

10.2.6 Traversity, Clustering and Density

The *traversity* or *clustering* is a property of a node in a network.[11] The *clustering coefficient* C is the mean probability that nodes sharing a common node are as well neighbors (Watts and Strogatz 1998; Scott 2000; Fronczak et al. 2002). It is assumed that if node A is linked to node B and node B to node C, there is a high probability

[10] Some authors use this term also for the average geodesic distance in a graph, although strictly the two quantities are quite distinct.

[11] It is important to note that the term clustering is probably misleading as it has also another meaning. A traditional method for extracting community structure from a network is called cluster analysis with a different connotation (Everitt 1974).

that node A is also linked to node C. In other words, the idea of a neighborhood is useful for the clustering coefficient of a network. It measures the cliquishness of a friendship network (Watts 1999). Accordingly, the clustering coefficient states how well the neighborhood of a node is connected. If it is fully connected, the clustering coefficient is 1, whereas a value close to 0 implies that there are hardly any connections. From a social network perspective, a high clustering coefficient implies that the friend of your friend is also likely to be your friend (Newman 2003b). The clustering coefficient, representing the density of triangles in a network, is defined locally and globally.

(a) *Local Clustering Coefficient.* A *local clustering coefficient* C_{li} is the density of the network at the local node i. More formally, it is equal to

$$C_{li} = \frac{\text{number of triangles connected to node } i}{\text{number of triples centered on node}}, \tag{10.4}$$

where a connected triple defines a node with edges to an unordered pair of other nodes (Watts and Strogatz 1998).

(b) *Global Clustering Coefficient.* The *global clustering coefficient* C_{gi} is the average clustering coefficient of all nodes in that graph, such that

$$C_{gi} = \frac{1}{n} \sum_i C_i, \tag{10.5}$$

if C_i is assumed to be 0 for nodes with degree 0 or 1.

It is interesting to note that sparse random graphs have a vanishingly small clustering coefficient, while real world networks frequently have large coefficients (Albert 2001).

10.2.7 Assortative Mixing Patterns

The identification of assortative mixing patterns in networks describes the patterns the connection of nodes have with each other. *Assortative mixing patterns* occur if the nodes in a network tend to be connected to other nodes with similar characteristics. Assortative mixing is quantified with an assortativity coefficient, such that the conditional probability $P(j \| i)$ represents the probability that node j has similar characteristics than the analyzed agent i. This coefficient is normalized between zero and one. Zero indicates a randomly mixed network, whereas one implies a perfect preferential assortation. Social networks are often assortatively mixed, while technological and biological networks tend to be disassortative. This property is important for the analysis of adoption and diffusion processes in software markets, as assortative networks tend to percolate more easily than their disassortative counterparts. Moreover, the assortative networks are also more robust to node removal, i.e. they have a higher network resilience (Newman 2002).

10.2.8 Degree Correlations

The *degree correlation* in a network is a particular case of assortative mixing, since it describes the mixing of nodes according to their node degree (Maslov and Sneppen 2002). In other words, the degree correlation investigates whether high-degree nodes of a network preferentially associate with other high-degree or with low-degree nodes. Degree correlations are of particular interest in network research as the degree of a node is itself a vital property of network topologies. Hence, degree correlations can give rise to some interesting network structure effects. There are several different ways to quantify and to display degree correlations. An intuitive approach is to visualize it as a two-dimensional histogram of the degrees of nodes at either ends of an edge (Maslov and Sneppen 2002). A more compact representation of degree correlations can be achieved by calculating the mean degree of the network neighbors of a node as a function of the degree k of that node. The resulting one-parameter curve increases with k if the network is assortatively mixed, e.g. the Internet (Pastor-Satorras et al. 2001). Moreover, it is possible to reduce the measurement to a single number by calculating the Pearson correlation coefficient of the degrees at either ends of an edge (Newman 2003a). The number is positive for assortatively mixed networks and negative for disassortative graphs. It is interesting to note that social networks tend to be assortative, with a high internal density between clustered edges and a low density among other clusters, whereas technical and biological networks appear to be rather disassortative.

10.2.9 Giant Component

The *component* of a network to which a node belongs is the set of nodes that can be reached from it by paths running along other links. In a directed graph a node has both an in-component and an out-component. These are the sets of nodes from which the node can be reached and which can be reached from it. Hence, the *giant component* of a network is the size of its biggest cluster[12] (Cohen et al. 2000; Broder et al. 2000; Callaway et al. 2001; Cohen et al. 2001; Dorogovtsev et al. 2001). It is an efficiency measure of the adoption and diffusion process. Sometimes the size of the second-largest component is also measured as the largest component is expected to be much larger than the second largest in networks well above the density at which

[12] The largest component is frequently equated with the giant component, although both are only the same in the limit of very large networks.

a giant component emerges. In software markets, the size of the largest component represents the largest fraction of customers within a compatible standard. Hence, the formation process of a giant component can be applied in order to analyze the critical mass of a network.

10.3 Prototypical Network Topologies

Based on the previously defined properties of networks, research discovered a variety of network topologies that are investigated in complex networks research. A network topology is the study of the arrangement or mapping of the elements of a network, e.g. the physical and logical links between nodes. More general, the topology of a network is defined based on any set X and a family T of subsets of X. Accordingly, T is termed a *topology* on X if:

1. The empty set and X are elements of T.
2. Any union of arbitrarily many elements of T is an element of T.
3. Any intersection of finitely many elements of T is an element of T.

Three robust measures of the network topology are the average path length, the clustering coefficient and the degree distribution of the network. (Albert 2001) Based on these measures, research on large-scale networks revealed three main classes of networks, i.e. topologies of networks, are particular relevant to the outlined research questions:[13]

1. *Random Graph Networks (RN).* Random graphs are based on a random distribution of links. They are frequently applied as a benchmark for computational or empirical studies.
2. *Small-World Networks (SWN).* Second, based on the discovery of clustering in networks a class of models has emerged which is collectively called small-world network models.
3. *Scale-Free Networks (SFN).* Finally, the discovery of powerlaw degree distributions in networks initiated various scale-free models that, by focusing on the network dynamics, aim to explain the origin of the power-law tails.

10.3.1 Random Graph Networks

Random networks are characterized by links which randomly connect to a number of edges (Karonski 1982; Bollobas 1985; Janson et al. 1999; Bollobas 2001).

[13] There is much more that could be said about prototypical network topologies, but a complete survey of all the material is beyond the scope of this paper. Please confer Barabasi (2002) and Newman et al. (2006) for further details.

They are characterized by a binomial or a Poisson degree distribution and can be visualized as circles with periodic boundary conditions.[14] Random networks are the oldest topology of networks and influenced research for decades after its introduction (Solomonoff and Rapoport 1951; Erdos and Renyi 1959). Today, researchers often concentrate on the limit behavior of random graphs, with an equal probability of edges. However, real world networks are frequently right-skewed with a long right tail of values (Broder et al. 2000). Nevertheless, random graphs are widely used for proving the existence of properties on network topologies.[15]

10.3.2 Small-World Networks

Networks are coined *small-world networks* in reminiscence to the small-world phenomenon which is also known as the "six degrees of separation". The small-world experiment was first described by the Hungarian Karinthy in 1929, and tested experimentally by Milgram and Travers (Milgram 1967; Travers and Milgram 1969). In essence, the experiment revealed that two arbitrary people in a social network are connected by only six degrees of separation, i.e. the diameter of the corresponding graph of social connections is not much larger than six. In other words, any individual in the world can reach any other individual through a very short chain of social ties. This was discovered by network researchers investigating the distribution of path lengths in social networks by asking participants to send a note to their close friends in order to reach a target person (Milgram 1967; Travers and Milgram 1969; Garfield 1979). The experiment revealed that the notes passed on average through six contacts, which is also termed six degrees of separation (Guare 1990). While the exact number may not be a very reliable estimate, additional research supports this insight that for some complex social networks only a short path is needed to connect even the most distant members (Sattenspiel and Simon 1988). Such networks may have a geographical component, as the nodes of the network have positions in space. And in many cases it is reasonable to assume that geographical proximity could play a role in deciding which nodes or agents of a network are connected to each others.

In 1998 Watts and Strogatz published a seminal paper with the first small-world network model (Watts and Strogatz 1998; Watts 1999). Their model revealed that by adding only a few long-range links, a regular graph, with a network diameter proportional to the network size, can be transformed into a small-world. In such a *small-world* the average number of edges between any two nodes is very small, whereas its clustering coefficient remains large. From a complex networks

[14] Please note that from a mathematical topological perspective a circle is a one-dimensional lattice.

[15] The Szemerédi regularity lemma states that if a property exists on a random graph this property exists on nearly all other network topologies. It is a derivative of the observation that as every large finite undirected graph can be approximated by a set of structured and pseudo-random parts (Komlós et al. 2002).

perspective, the characteristic properties of small-world networks are an asymmetric development of their characteristic path length and their cluster coefficient at changing rewiring probabilities. The *Small-World Network Analysis* depicted in Fig. 12.5 reveals this relationship.[16] While the global clustering coefficient C_{gi} decreases with increasing rewiring probabilities, the diameter decreases exponenentially. Essentially, the key insight of this phenomenon is the small number of short-cuts that are necessary to design a small-world network, and the difficulty for an observer to notice this (Buchanan 2002). In other words, if all nodes are connected to each other by relatively few edges, even a few short-cuts imply that the average path lengths of the network decreases exponentially. This implies that small-world networks are cliquish (Watts and Strogatz 1998). Small-world properties are useful for determining a variety of applications, e.g. to determine efficient distributions of edges for optimal information and control flows of networks. Once the required normalized network measures the average path lengths, and the cluster coefficient are calculated for different relationship thresholds, a small-world test can be performed and its results can be compared to real-world networks and other small-world networks. While the significance of this research is largely proven in natural scientific research areas, its application in social sciences are recent phenomena. A review of these successful applications reveals that the two most important generation algorithms for small-world networks are rewiring and the shortcut method (Watts and Strogatz 1998; Watts 1999).

(a) *Rewiring Method.* According to the rewiring method, one end of each link of a regular network is reconnected with the probability p to a new location that is chosen uniformly at random from the lattice.[17] The respective rewiring probability p has to be chosen such that there are a low density of shortcuts in order to observe a high degree of clustering and a small diameter. If p is close to unity, the resulting network will resemble a random network, whereas $p = 0$ in a regular lattice. Between both extremes exists a sizable region for which the model has both low path lengths in combination with a high transitivity.
(b) *Shortcut Method.* In an alternative variant, no edges are rewired. Instead, shortcuts of joining randomly chosen pairs of nodes are added to a low-dimensional lattice (Newman and Watts 1999). Accordingly, a new parameter p is defined that governs the density of these shortcuts. It represents the probability per edge of there being a shortcut anywhere in the network. Hence, the mean total number of shortcuts is Lkp, while the mean degree is $2Lk(1 + p)$. This alternative approach has the advantage that no edge ever become disconnected from the

[16] Please note that in a Small-World Network Analysis the development of the clustering coefficient of a network is compared to the development of its characteristic path lengths for varying rewiring probabilities β (Watts 1999). Moreover, it is important to note that such investigations are not limited to small-world graphs, but can be conducted also for other network topologies in order to investigate vital differences among network topologies.

[17] Please note that the rewiring probability is also termed *beta*. This is the name of the respective variable in the complex networks adoption and diffusion simulator.

rest of the network, implying that the mean node distance is always formally finite (Newman 2000).[18]

10.3.3 Scale-Free Networks

Scale-free networks are characterized by a powerlaw degree distribution which has been found in many contexts, such as citation networks among scientific papers, the World Wide Web, the Internet represented by physical routers, or e-mail graphs (Price 1965; Barabasi et al. 1999; Albert et al. 1999; Faloutsos et al. 1999; Strogatz 2001; Ebel et al. 2002; Albert and Barabasi 2002; Dorogovtsev and Mendes 2002).[19] The complex networks research on scale-free networks dates back to the late 1990s with the discovery of a powerlaw degree distribution in many real world networks. Although many are ambiguous power laws, the broad spectrum, both in degree and in domain, reveals that scale-free networks are clearly very different from other network topologies. Thus, some nodes in scale-free networks have a degree that is orders of magnitude larger than the average. Such nodes are frequently called hubs.[20] Similar functional forms for the degree distribution are powerlaws with exponential cutoffs, e.g. such as in the network of movie actors or collaboration networks (Newman 2001b). Moreover, it is important to note that while a particular form may be seen in the degree distribution for the network as a whole, subnetworks of the network can have other forms. An example is the World Wide Web that exhibits a powerlaw degree distribution overall, although individual domains reveal unimodal distributions (Pennock et al. 2002). There are a variety of approaches to construct scale-free networks. The Yule process is a generic process to create networks with the characteristic powerlaw degree distribution that has been recognized since 1925 (Yule 1925; Simon 1955).[21] However, this approach is also known by many other names such as the Matthew effect, cumulative advantage and, most recently, preferential attachment, which are all based on the underlying paradigm that existing links attract additional links (Barabasi et al. 1999).[22] The most recent mechanism for the generation of scale-free networks is the preferential attachment

[18] Please confer (Newman 2000) for further mathematical or physical details on both methods.

[19] In other words, a network is termed *scale-free* if the probability that a node selected uniformly at random has a certain number of links follows a mathematical function called a powerlaw. The origin of the term refers to any functional form $f(x)$ that is unchanged to within a multiplicative factor under a rescaling of the independent variable x. In essence, this implies powerlaw forms, which are the sole solutions to $f(ax) = bf(x)$. For this reason, the terms powerlaw and scale-free are used as synonyms in the following.

[20] Please note that this expression is misleading as there is no inherent threshold above which a node can be viewed as a *hub*. Otherwise, the network would not have a scale-free distribution.

[21] In statistics, the YuleŰSimon distribution is a discrete probability distribution named after Udny Yule and Herbert Simon which is the result of this Yule process.

[22] Please confer also the Sect. 10.4.

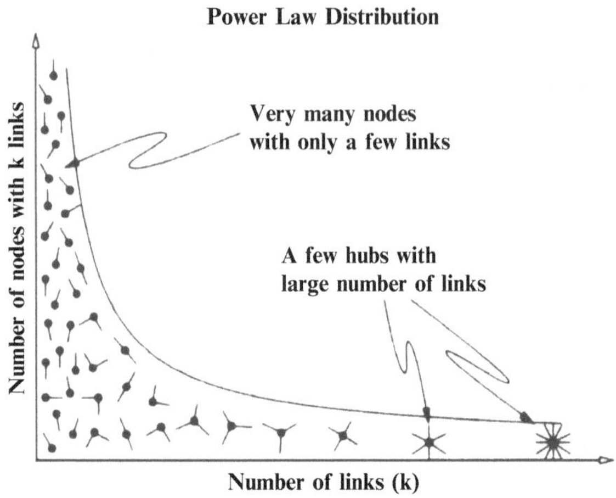

Fig. 10.1 Scale-free network distribution
Source: Barabasi (2002)

of Barabasi and Albert (Barabasi et al. 1999). Accordingly, a scale-free network is constructed by starting with initial nodes and then keeps adding additional nodes and links with a probability that is proportional to the degree of the already existing nodes (Fig. 10.1). [23]

10.4 Evolution of Complex Networks

In order to design network models representing customer networks of software markets it is necessary to understand the *evolution of networks*, describing the network assembly and their development (Barabasi et al. 1999). While the goal of the previously summarized properties and topologies was to construct networks with topologies similar to real world networks, the focus of this chapter is to capture their evolution, i.e. their emergence and development. But progress on this research frontier has been slower than understanding static network properties. Nevertheless, there are some important recent advances, particularly with respect to network growth processes. All of the models discussed so far concern static properties of networks that allow one to create the respective network topologies, but they do not contribute to the understanding of how and why networks come to have these properties. In general, the networks are typically assumed to grow by the addition of links and edges in some manner intended to reflect growth processes (Banks and Carley 1996). In this section first the archetypal cumulative advantage model of Price is outlined, which refers back to work by Simon. Then, the highly influential

[23] This mechanism is sometimes also termed "the rich getting richer".

preferential attachment model and a number of variations and generalizations are depicted.[24]

10.4.1 Cumulative Advantage Model

Based on the pioneering work of Simon, Price described in 1965 with his *cumulative advantage model*, which is in all likelihood the first example of a scale-free network, when he investigated the citation network of researchers. He discovered that both in- and outdegrees of academic authors follow a powerlaw distribution. (Simon 1962; Price 1965) A few years later he offered an accepted explanation for the growth processes of such powerlaw degree distributions (Price 1976). Herbert Simon showed that power laws arise when "the rich get richer", which in sociology is referred to as the "Matthew effect" after a biblical edict (Simon 1962; Price 1965; Merton 1968). Price calls this phenomenon cumulative advantage (Price 1965). Despite the similarities there are important differences among the research contributions. Simon was investigating wealth distributions, whereas the important contribution from Price was to apply the ideas of Simon to the growth of a network (Price 1965). Price was the first to discuss cumulative advantage specifically in the context of networks based on the underlying idea that the rate at which a paper is cited should be proportional to the number of citations that it already has. This is plausible as the probability that one comes across a specific paper increases with the number of other papers that cite it. For this reason, the probability that a paper is cited increases at a similar rate. Price used a master-equation or rate-equation method. Unlike today, he did not have computational resources to develop simulations of the model, and so he could not provide numerical results. But the underlying rationale can be also transferred to other networks. By studying other real-world networks a great deal more progress has been recently made in understanding cumulative advantage processes and the growth of networks. Hence, it is reasonable to consider the cumulative advantage model for an implementation in a complex networks software market model. But most of the recent contributions have been carried out using a slightly different model. Today, this phenomenon is usually known as preferential attachment model.

10.4.2 Preferential Attachment Model

The cumulative advantage idea did not achieve much attention until its rediscovery some decades later by Barabasi and Albert. They suggested a *preferential*

[24] There is much more that could be said about the evolution of complex networks, but a complete survey of all the material is beyond the scope of this paper. Please confer Barabasi (2002) and Newman (2003b) for further details.

attachment model for the Web that is very similar to the outlined cumulative advantage model, but with an important difference (Barabasi and Albert 1999). Both models are similar as they add edges to the network with degree m, which is never changed thereafter. The other end of each edge is connected to other edges with a probability that is proportional to the degree of that node. A decisive difference between the models concerns the degrees of the edges as the edges are undirected in the preferential attachment model, i.e. there is no distinction between in- and outdegrees of edges. This difference has advantages and disadvantages (Newman 2003b). While in reality many networks are directed graphs, by ignoring the directed nature of the network, the preferential attachment model avoids the problem of how an edge gets its first link as each node has an initial degree m, and thereby automatically a non-zero probability of receiving new links. Consequently, the model sacrifices some of the realism of the cumulative advantage model in favor of simplicity.

But in addition, some other insights can also be derived from the preferential attachment model. The model suggests a correlation between the age of edges and their respective degrees (Adamic and Huberman 2000). As older nodes tend to have higher mean degrees, the earliest edges added to the network have substantially higher expected degrees than those added later. This implies that the overall powerlaw degree distribution of the whole graph is a result primarily of the earliest edges. Applied to customer networks in software markets this implies that early customers are particularly important, as they are likely to be the most connected in the emerging network. In turn, this correlation has been used to criticize the preferential attachment model. Investigations revealed that there are no such correlations in the real networks, but this finding does not necessarily imply that preferential attachment is not the explanation for powerlaw degree distributions in such webs (Barabasi et al. 2000; Adamic and Huberman 2000; Bianconi and Barabasi 2001b; Bianconi and Barabasi 2001a). Moreover, the preferential attachment model is in the original version a model of an undirected network. It can be extended to a model of a directed network, but then attachment is proportional to the sum of in and out-degrees, which is unrealistic as it should be in proportion to in-degree only. In addition, if the preferential attachment model is applied to directed network, it generates acyclic graphs, which are a poor representation of real world networks. Finally, the nodes belong to a single connected component, while in the real Web there are many separate components and strongly connected components.

In order to account for this criticism, many extensions and modifications have been suggested in order to make it more realistic to real world networks. The most relevant generalizations to the design of customer networks for valuation in software markets are presented in this subsection.[25]

Nonlinear Attachment Probability. The first generalization assumes that the attachment probability of a node is not linear to its degree k, but follows a *more general power of degree* (Krapivsky et al. 2001). This generalization implies

[25] Please confer (Albert et al. 1999) for further information on the generalization of the preferential attachment model.

three possible classes of behavior depending on the exact degree of the attachment probability: For $\gamma = 1$, the normal linear preferential attachment is derived, while for $\gamma < 1$, the degree distribution is a powerlaw multiplied by a stretched exponential. For $\gamma > 1$ follows a condensation phenomenon, i.e. a single nodes receives a finite fraction of all the connections in the network. Finally, for $\gamma > 2$ follows a non-zero probability that the node will be connected to every other node.

Change of Mean Degree. Another variation is to *change the mean degree over time*, i.e. increase of parameter m over time (Dorogovtsev and Mendes 2001). Hence, it is assumed that the number of new edges m added per new edge increases with network size n as n^a for some constant a, and that the probability of attaching to a given edge is $k + Bna$ for a constant B. Hence, the resulting degree distribution follows a powerlaw with exponent.

Addition of Nodes and Links. In the model edges are not added once they exist, but instead remain where they are. While this assumption is reasonable for some networks, such as citation networks, it is not realistic for others, e.g. the World Wide Web. For this reason, some researchers have developed a model that adds to the standard preferential attachment model an extra mechanism for node appearance and disappearance based on stochastic constant but with different rates. Investigations reveal that over a wide range of values, the powerlaw degree distribution is maintained.

The investigations reveal that the original preferential attachment model is elegant and simple, but lacks a number of features that are present in real world networks. However it is important to note that a variety of generalizations and modifications of the model are available. Hence, it could be possible to apply extended versions in order to model the evolution of complex customer networks in software markets.

10.5 Processes on Complex Networks

The focus of this section is on processes taking place on complex networks, such as product diffusions on customer networks in software markets. The ultimate goal is to extract relevant information from the networks in order to enhance the valuation of software companies. Thus, the next logical step is to investigate the most relevant processes, e.g. epidemiological processes, percolation, and phase transitions.[26] Based on these insights, a complex networks adoption and diffusion simulator is

[26] Please note that these concepts are not mutually exclusive and collectively exhaustive as they can complement each other in the investigations of diffusion dynamics, e.g. it is possible to investigate phase transitions of a diffusion process which is modeled with an SIR model.

designed in order to investigate customer networks for a customer network-centric valuation in software markets.[27]

10.5.1 Epidemiology

Epidemiology is highly relevant to the research on adoption and diffusion processes in software markets as its main research goal is to understand the mechanisms by which information, products, resources, and contagious diseases spread on various networks (Klovdahl et al. 1994; Ebel et al. 2002; Newman et al. 2002). Depending on the complexity of the underlying assumptions a broad spectrum of epidemiological endemic and epidemic models are available. The most relevant models are the SIR model and the related SIS model. The SIR model is a model of *epidemic diseases* such as influenza, which infect a significant fraction of individuals in a short outbreak. The SIS model, in turn, is a model of *endemic diseases* such as measles that persist within the population at a level roughly constant over time.

10.5.1.1 SIR Model

The standard epidemiological model describing the propagation of contagious diseases in social networks is the *Susceptibles, Infective, Recovered (SIR) - Model* (McKendrick 1926). According to this model, the population can be classified into three strata comprising susceptibles (S), infectives (I), and recovered (R).[28] *Susceptibles* are individuals that do not have a disease, yet. But they could catch it if they are exposed to an infective. *Infectives* have the disease and can pass it on to susceptibles. *Recovered* individuals have recovered from the disease and are assumed to have permanent immunity.[29] Depending on the infection rate, the characteristics of the interacting agents, and on the network configuration, it is possible to investigate the characteristic dynamics of a disease in a specific network with analytical and numerical models (Bailey 1975; Anderson and May 1991; Newman 2003b).

1. *Analytical Differential Equations.* A classical approach to model the epidemiological dynamics analytically, is to assume that susceptibles have a uniform

[27] Please note that there is much more that could be said about processes on networks, but a complete survey of all the material is beyond the scope of this paper. Please confer (Strogatz 2001; Albert 2001; Newman 2003b; Newman et al. 2006; Kemper 2006) for further details on processes on complex networks.

[28] Please note that although the more common word is infectious the standard term among epidemiologists is infective.

[29] While some authors also use the term removed implying the possibility that people may die of the disease and are removed from the infective pool, others, studying reaction diffusion processes use the term refractory (Strogatz 1994).

probability per unit time to get the disease from any infective. In addition, infectives may recover at some stochastically constant rate. Hence, the respective fractions s, i and r of individuals in the states S, I and R are determined with differential equations (Hethcote 2000). An analytical methodology for solving the respective differential equations is the mean-field approximation. This assumes that the influence on an individual is equal to the average effect of all short-range neighbors (Barabasi et al. 1999; Pastor-Satorras and Vespignani 2001a). This is called a *fully mixed model*, which reveals much about the basic dynamics of diseases. However, it is important to note that this fully mixed model is based, as all analytical approaches, on the assumption that all individuals of the population are connected. This is obviously unrealistic as in reality diseases can only spread between those individuals who have actual physical contact. Therefore, such analytical approximations illuminate all topological information about customer networks.
2. *Percolation Models.* Alternatively, the epidemiological diffusion dynamics can be investigated with percolation models.[30] Hence, the model is generalized, although the resulting dynamical system is substantially more complicated. The underlying observation of the generalization is that the model can be mapped onto a bond percolation model (Grassberger 1983; Sander et al. 2002). According to this approach the distribution of percolation clusters corresponds to the distribution of disease outbreaks. In addition, the percolation transition can be interpreted as the epidemic threshold above which an epidemic outbreak is possible. And the size of the giant component represents the size of the epidemic.[31] A disadvantage of the percolation model is that the time progression of a disease outbreak is not available as it only provides results in the limit of long time horizons. Nevertheless, there is much to be learned by studying the bond percolation model.
3. *Numerical Simulations.* The diffusion dynamics of epidemics can be also studied with numerical simulations which allow to increase the level of detail. Applied to software markets all potential customers are simulated and the infection rate represents the probability that a customer will purchase a product due to the impulse of an infected neighbor. Hence, the infection rate is a vital parameter influencing the spreading behavior of a diffusion process. It represents the susceptibility to infection, and is determined by dividing the adoption rate by the recovery rate. In the SIR model, the recovery rate is the counterbalance to the diffusion rate that represents the probability that customers stop to use the software product, which is also expressed as a probability in percentage terms. Vital determinants influencing this recovery rate are competition, product life cycle, and changes in personal and technological preferences. Since the purchase decision is influenced by many factors such as the marketing strategy, the need for the product,

---

[30] Please confer Sect. 10.5.2 for further details.

[31] The epidemic threshold exists if a non-zero fraction of the population is infected in the limit of large networks.

and other market parameters, the infection rate has to be adjusted with respect to the relevant aspects. In contrast, the recovery rate is equal to the sum of effects that might hamper consumption, e.g. force of substitution, competition, and others. Once the model is designed, multiple iterations of the simulation reveal the dynamics of the network diffusion process as the respective outcomes can be depicted according to their probabilistic distribution.

10.5.1.2 SIS Model

In contrast to the epidemic SIR model, processes on networks can also be endemic, i.e. they are not self-limiting and can infect susceptibles several times. Hence, recovered carriers of the product are considered as part of the susceptible pool, again. This implies that the diffusion process could persist indefinitely, as it could repeatedly circulating in the population. This model is termed the *SIS model*. Illustrative examples are computer viruses that can be removed by antivirus software, but the virus can infect the computer, again. This epidemiological model can be applied to product sales in software markets, as software can be bought multiple times. The diffusion process is governed similarly to the SIR epidemic transition by a phase boundary with parameter regimes in which the disease persists, and those in which it does not. The SIS model cannot be exactly analytically solved on a network, but it can be approximated by a mean-field approximation (Pastor-Satorras and Vespignani 2001a).

Both approaches, the SIR and the SIS model, allow an investigation of the diffusion dynamics in customer networks software. If the required parameters are available, tested and challenged in a sensitivity analysis, a mean expectation and the corresponding volatility of the diffusion can be derived. Moreover, potential phase transition boundaries and the probabilities for crossing them can be revealed, and the range of the involved uncertainty is illustrated. Thereby, the models contributes to a better understanding of diffusion dynamics in customer networks of companies operating in software markets.

10.5.2 Percolation and Phase Transitions

Additional insights into dynamics of complex networks can be derived by considering percolation processes and phase transitions. The *percolation theory* is a concept of condensed-matter physics for the description of random physical processes, such as flows through a disordered porous medium (Grassberger 1983; Bollobas and Riordan 2006). Originally, the percolation theory was applied in order to examine natural scientific percolations. For example, electric flows in electronic circuits reveal a sharp transition at which a long-range connectivity appears if edges are randomly added. It dates back to the 1950s, when percolation models were applied in order to model the spread of diseases (Broadbent and Hammersley 1957).

Depending on the porous medium, it is necessary to distinguish bond percolation and site percolation models. In site percolation nodes, which are also called sites, are either occupied or unoccupied and valves are placed at the intersections, rather than in the pipes of the network. The research focus is on the shape and size of the emerging clusters of occupied sites. In bond percolation, in turn, edges, which are also called bonds, are investigated.

Similarly, the models can be applied to networks in order to study the dynamic behavior of a system by selecting different state variables at random edges and nodes (Schwartz et al. 2002). Both percolation models allow one to identify sharp transitions, i.e. phase transitions, at which the long-range connectivity of a network appears or disappears if edges are randomly added or deleted, respectively. A *phase transition* is a concept of thermodynamics. It describes the transformation of a system from one phase to another at a critical phase transition point, where physical properties undergo abrupt changes, e.g. the transition of liquid water into vapour at boiling point (Freeman 1977; Bar-Yam 1997; Jeong et al. 2001; Luterotti and Stefanelli 2002). In other words, a phase transition is a sudden switch of system behavior at a critical parameter value. The switch occurs at the macroscopic level result from behavior its elements at the microscopic level. The identification of such phase transitions in percolation models allows one to determine epidemic thresholds of epidemic outbreaks in complex networks (Hethcote 2000). Hence, they are complementary to the outlined epidemiological approaches (Fig. 10.2).

Applied to complex customer networks in software markets the focus is on various properties of nodes and links in networks, which are randomly designated either occupied or unoccupied. Accordingly, a *critical mass of customers* can be interpreted as the critical parameter of a phase transition in the product diffusion

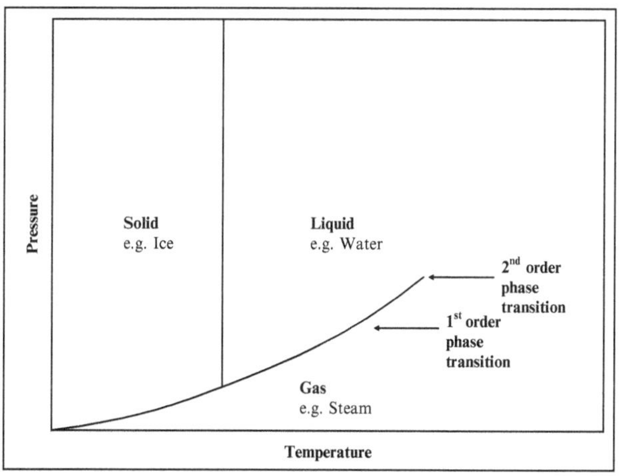

Fig. 10.2 Schematic phase diagram of H_2O
Source: Bar-Yam (1997)

process.[32] This allows to determine and quantify the critical mass of customers exactly at the emergence of a giant component in customer networks. Hence, such investigations of phase transitions in percolation models are suitable concepts for the analysis of product diffusion processes in software markets.

10.6 Derivation of Complex Networks Research Hypotheses

All in all, the previous chapter reveals that the complex networks theory has a considerable explanatory potential with respect to some of the open research questions on customer network-centric valuation in software markets. Based on the fundamental insights on properties, topologies and dynamics of networks, the open research questions are approached by deriving complex networks research hypotheses which are investigated in the rest of this part. The goal of these complex networks investigations is to derive additional insights into complex customer networks based on which the network effects framework can be enhanced. For this purpose, the following hypotheses are particularly relevant as depicted in their respective motivation.

1. *Diffusion Dynamics in Varying Network Topologies.* Complex networks research identified a variety of topologies among networks (Watts 1999; Albert 2001; Barabasi et al. 2000). Based on these findings we formulate the hypothesis that the variation of customer network topologies has an impact on the adoption and diffusion behavior in software markets. [HP1] This hypothesis is a central link between the outlined research on valuation, network effects and complex networks. It is motivated by the complex networks finding that there are, in general, various network topologies with different adoption and diffusion dynamics. Applied to customer networks in software markets this would indicate that the topology of customer networks is relevant in the analysis of customer network developments. These, in turn, are decisive for customer network-centric valuation in software markets.
2. *Scaling Properties of Complex Customer Networks.* If the topology of complex customer networks is relevant, complex networks research could be relevant to derive additional insights. But full-size models of large-scale networks are frequently not possible due to computational limitations. For this reason, a prerequisite for modeling is that the relevant properties and dynamics of networks are invariant to scaling. Initial research efforts indicate that the scale of the network does not have an impact on its properties and dynamics, but the respective hypothesis are not proven (Weitzel et al. 2000; von Westarp 2003). Based on this research evidence, we formulate the hypothesis that relevant properties and dynamics of customer networks in software markets are invariant to scaling. [HP2] The motivation for this hypothesis is to investigate the customer networks

[32] Please note that this critical mass is different from the definition of an economic break-even of a company.

by upscaling, e.g. to increase the size of the customer networks in order to study their growth processes, as well as by downscaling, i.e. creating size-reduced models of large scale real-world networks.

3. *Network Topologies of Complex Customer Networks.* If network topologies are relevant it would be interesting to specify the exact nature of customer networks in software markets. Complex networks research reveals that most real world social networks are small-world networks. (Watts 1999; Albert 2001; Barabasi et al. 2000) Hence, we frame the hypothesis that customer networks in software markets also tend to be small-world networks with the respective properties and dynamic of networks in this equivalence class. [HP3] This hypothesis is of primary interest for the open research questions as it would allow one to transfer network theoretical insights from small-world networks to complex customer networks in software markets.
4. *Contributions to Customer Network-Centric Valuation.* If the previously formulated research questions are resolved, it would be interesting to apply the complex networks insights in order to enhance customer network-centric valuation in software markets. Previous investigations of the relevant valuation approaches emphasized the vital importance of the price of the underlying and of the volatility as parameters in the valuation process. If customer networks are interpreted as the underlying of a software company, the complex networks investigations of the customer networks may contain additional information that could be applied in order to better derive approximations of the input parameters and thereby to increase the quality of the valuation. Hence, we derive the hypothesis that the complex networks tools can be applied to customer networks in software markets in order to derive supportive information on valuation parameters if the required information on the underlying customer network is accessible and reliable. [HP4] This hypothesis is the fundamental link between the previous research on corporate valuation in software markets, network effects and the outlined complex networks research.
5. *Social versus Natural Scientific Networks.* If complex networks research is applied to customer network-centric valuation in software markets, it is also necessary to be aware of the underlying assumptions and the respective limitations of this interdisciplinary research approach in the context of the given research. Complex networks theory reveals a variety of universal properties and dynamics of networks (Newman 2003b; Newman et al. 2006). Nevertheless, we hypothesize that customer networks in software markets are peculiar due to their social nature which is different from electronic, information or other scientific networks. [HP5] This hypothesis is motivated by shifting the focus of the research to the limitations of the interdisciplinary approach.

The following investigations are intended to fill the identified research gaps on valuation in software markets. Therefore, the derived hypotheses are investigated with an adoption and diffusion simulator and complementary complex networks research in a complex networks analysis of customer networks. But previously, the required complex networks adoption and diffusion simulator is designed in the next chapter.

Chapter 11
Complex Networks Adoption and Diffusion Simulator

"The dynamic spatial redistribution of individuals is a key driving force of various spatiotemporal phenomena on geographical scales."

D. Bernoulli (1760)

A review of the complex networks theory reveals a variety of important insights into customer-centric valuation in software markets. These allow one the formulation of the research hypotheses concerning the properties, dynamics and topologies of customer networks in software markets. In this chapter, a numerical complex networks adoption and diffusion simulator is developed for a two-fold purpose. First, the simulator is designed, as stated, in order to investigate the hypotheses in the following complex networks analysis of customer networks. The second, more general motivation is to provide a guideline for the design of a simulator that can be applied in order to investigate complex customer networks of real world software companies. Therefore, it is integrated in a later chapter of the book into the previously developed network effects framework. The result is a complex networks framework for valuation in software markets based on the complex networks adoption and diffusion simulator. For both reasons, the purpose of this chapter is to provide an overview of the design and implementation process as well as on the main features of the simulator.[1]

11.1 Object-Oriented Software Engineering

This section summarizes the most important information on the design approach of the complex networks diffusion simulator.[2] The development of the simulator follows the object-oriented software engineering paradigm, which allows modular

[1] Please note that this chapter only summarizes the most relevant mechanics of the program which are required in order to understand the subsequent analyzes.

[2] The term *simulation* is defined as controlled experiments with or on methods in order to mimic real world phenomena (Meissner 1970). Please confer (Meissner 1970) for further methodological details.

A. Kemper, *Valuation of Network Effects in Software Markets*,
Contributions to Management Science, DOI 10.1007/978-3-7908-2367-7_11,

modifications.[3] Accordingly, the development of the program can be separated into three different but interdependent steps: (Booch 2007)

1. *Object-Oriented Requirements Analysis*. First, an object-oriented analysis is conducted in order to determine the design requirements of the program. The goal of this first phase is to understand from a programming design perspective what the program is supposed to do. For this purpose, object-modeling techniques are applied in order to analyze the functional requirements
2. *Object-Oriented Design*. In the second object-oriented design phase, a specification of the previous analysis is developed based on programming design patterns. The aim is to interpret the previous functional analysis and to optimize the implementation of the program. Hence, the focus of this phase is on how the system should work.
3. *Object-Oriented Implementation*. The purpose of the final object-oriented implementation phase is to transform the design specifications into a software code. The focus is to implement a program with an optimal performance that provides the previously defined features based on the identified programming design patterns.

11.2 Object-Oriented Requirements Analysis

The overall goal of the simulator is to investigate the properties, topologies and dynamics of complex customer networks in software markets. Hence, it is required to provide the following four generic functions:

1. *Network Generation*. First, it is necessary that the program allows one to create networks representing the customer network in software markets. In particular, the simulator should be capable to generate the previously defined relevant network topologies.
2. *Customer Adoption Rule*. Another vital functionality of the program is to model the product diffusion process itself. Hence, the definition of a customer adoption rule is central to this phase. This rule determines the purchasing decisions of customers in the simulated software markets.
3. *Network Visualization*. In addition, the simulator should be capable to visualize the adoption and diffusion process as such a visualization illustrates the dynamics of the investigated complex customer networks. As this task requires considerable computational power even for small networks, it is reasonable to separate it from the computations of the adoption and diffusion process based on the

[3] *Object-orientation* is a software engineering approach that models a system as a group of interacting objects (Booch 2007). In such a system each object represents some relevant entity that is described by its class, its state, and its behavior. Object-oriented models can be created to show the static structure, dynamic behavior, and run-time deployment of these collaborating objects.

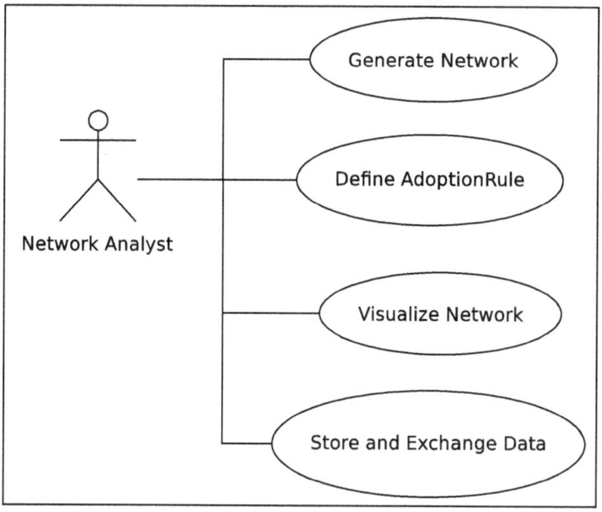

Fig. 11.1 Use case diagram
Source: Author

specified customer adoption rule. Moreover, it should be executed just for small to medium sized networks in order to limit the required computational resources.

4. *Data Storage and Exchange*. Finally, the program should be capable to record and to store the simulation results such that the respective network data can be analyzed. Furthermore, it is desirable that the program is capable of loading stored data, especially previously generated network instances. Therefore, an input and output data interface is required.

These requirements are visualized in the following Use Case Diagram (Fig. 11.1). The four main functionalities of the simulator are described in more detail in the following subsections.

11.2.1 Network Generation

In the first step, networks have to be generated that represent the network of actual and potential customers of the software. Due to their importance, the network generator should be capable of to generate the following most relevant topologies:[4]

1. *Random Networks*. Random networks are characterized by links which are connected randomly to a number of nodes (Karonski 1982; Janson et al. 1999). Hence, the simulator should allow one to generate a network with N nodes that have an average connectivity C, i.e., each node has on average C incoming links.

[4] Please confer also section 10.3.

2. *Small-World Networks*. Small-world networks interpolate between a random network and a lattice. While the average number of edges between any two nodes is very small, the clustering coefficient of the total network is large.
3. *Scale-Free Networks*. Scale-free networks are characterized by a powerlaw degree distribution, i.e. the probability that a node selected uniformly at random has a certain number of links follows a mathematical function called a powerlaw.

Once the relevant customer network topology is created, the simulator should permit the functionality to store it for further investigations. At the same time, this implies that it is also necessary to implement a feature for loading the complex customer networks once it was generated and stored.

11.2.2 Customer Adoption Rule

In the second step, the adoption decision that determines the purchasing behavior of the investigated population is codified. Based on the insights on purchasing decision models in software markets, the complex networks diffusion simulator could account for the following adoption and diffusion rules and features:

1. *Decentralized Standardization Model*. The decentralized adoption model compares the direct and indirect benefits and costs of the software.[5] Hence, the respective parameters are required in order to calculate the net benefits of each actor in each round.
2. *Mean-Field Approximation*. Alternatively, it is also possible to implement the adoption rule as a mean-field approximation. This mean-field approach approximates the many-agent system of independent agents as a field created by other agents through averaging over one representative neighborhood. In other words, the mean-field approximation simplifies the problem to the description of a representative field created by other members of the customer network. This field represents the simultaneous average opinion of all neighbors. The mean-field approximation is frequently implemented with a master-equation approach (Aoki 2005).

A comparison of both approaches reveals that the decenralized standardization model is a local interaction model, while the mean-field approximation is a long-range interaction model. A crucial assumption of the adoption rule of the decentralized standardization model is that every economic actor estimates that the adoption probability of his direct neighbors is similar to his own parameters. This implies an underlying local interaction rule as adoption decisions are primarily based on the relationship to the closest neighbors who have already adopted the software. The mean-field approximation, in turn, is more suitable if the interactions have primarily a long-range character, e.g. television or media marketing campaigns that

[5] Please confer section 7.2.2.2.

influence the adoption decisions of unrelated individuals. With respect to adoption and diffusion processes in software markets, the previous investigations revealed the predominant importance of local interactions in complex customer networks due to network effects, e.g. due to word-of-mouth referrals. Therefore, we focus our implementation efforts on the decentralized standardization model and extend its functionality for a stochastic description of benefits based on a standard normal distribution.

11.2.3 Network Visualization

Once a complex networks software market model is designed, the selected customer adoption rule is iterated. In order to deliver a visual impression of the investigated adoption and diffusion process, the simulator should provide a visual representation. This representation requires considerable computational resources that increase exponentially with the size of the simulated complex customer network. But visual impressions should be implemented at least for small networks in order to provide a visual intuition of the diffusion dynamics in the investigated software markets. The following features are particularly relevant for customer network-centric valuation in software markets:

1. *Bayesian Network*. The customer network could be depicted as a Bayesian network. It is a very intuitive but also a very computational intensive form to depict networks. The visual representation of dynamics in large-scale networks, e.g. networks with more than 100,000 nodes, is nontrivial and requires large computational resources. Despite of efficient network design and diffusion algorithms even high performance computers are not capable to calculate and visualize adoption and diffusion processes in large-scale networks. These problems are at the forefront of computer scientific research, e.g. visualization of large-scale networks, multigrid computing, or massively parallel processing (MPP). Therefore, the number of visualized nodes should be restricted. In the most recent version of the simulator the maximum size of the customer network visualization is limited to 1,000 nodes. With respect to the adoption and diffusion process, in turn, the most recent version was capable to compute adoption and diffusion dynamics in a random network with 300,000 nodes with an average degree of 5 or in a random network with 50,000 nodes with an average degree of 50. Please note that the maximum size depends also on the network topology as these are implemented with different network generation algorithms.
2. *Colored Diffusion*. The dynamics of the adoption process can be illustrated by coloring the nodes of the complex customer networks. Accordingly, each potential customers are colored in blue. They turn red as soon as the software is purchased. This coloring provides visual intuitive impressions of the product diffusion dynamics governing the complex customer networks in software markets.

11.2.4 Data Storage and Exchange

The simulator has to store the experimental data in a database for further investigations. Hence, the complex networks diffusion simulator should provide the following features:

Textual Output Table. The main data of the adoption process, e.g. the number of adopters, should be recorded in a textual table. This table should depict the status of the most relevant parameters in order to allow an efficient supervision of the adoption and diffusion process.

Excel CSV File. After each round the simulation results should be recorded in a CSV file. This allows also subsequent investigations with external software, e.g. MatLab, Microsoft Excel, SPSS or R.

11.2.5 Extended Program Features

The following additional features are useful extensions of the presented core functionalities.

1. *Network Theoretical Analysis*. In the core version of the simulator, the export of data allows further investigations with external programs. In addition, it would be desirable to investigate properties and dynamics of networks realtime, i.e. during the simulation. This should be an additional analytical feature of the simulator as it may require considerable computational resources. The computational disadvantage has to be traded against the possibility to get quickly an overview on the most relevant adoption and diffusion measures. These comprise e.g. the diameter, the giant component, and the clustering coefficient. Depending on the purpose of further investigations, the simulator should allow to add further network theoretical measures.
2. *Stochastic Utility Distribution*. It is difficult to measure and to quantify the utility of a software as the utility of two similar but not identical users is likely to be different, even for the same product. But the quantification of the benefits is a critical input parameter of the decentralized adoption model that has a considerable impact on the diffusion dynamics and on the outcome. Therefore, it should be possible, in a second step, to refine the adoption rule. For this purpose, a central observation is that it is frequently difficult to determine the exact benefit of a single economic agent, whereas it is often possible to describe the overall benefit distribution of a group. Hence, the simulator should provide an option to enter the benefits in the form of a benefit distribution in order to increase the realism of the model. Moreover, the assumption of the relevant benefit distribution should be capsulated in order to implement other distributions as well, if the respective

empirical data is available.[6] But in the first step it is reasonable to assume that the benefits follow a normal distribution from which the individual values are drawn.[7] This is a plausible assumption as software products are frequently perceived to be particularly beneficial or irrelevant for a minority of the population, while the majority attributes a medium level benefit. Accordingly, the simulator has to request the mean and the standard deviation of the assumed standard distribution.

11.3 Object-Oriented Design

After the definition of the requirements the program is designed.

11.3.1 Design Patterns

For this purpose, supportive programming design patterns are identified in order to optimize the performance of the implementation (Gamma et al. 1995). The following three design patterns are particularly important in the design of the adoption and diffusion simulator.

1. *Abstract Factory*. The creation of a network instance is designed as an abstract factory. This design pattern is an interface for creating families of related objects, i.e. customers and links between customers, without specifying their concrete classes. The motivation to apply the abstract factory concept is to abstract from concrete classes in order to maintain a maximum degree of flexibility for further modifications of the simulator.
2. *Decorator*. Nodes and edges are designed based on the decorator pattern. This decorator approach allows one to attach additional responsibilities to important objects of the program. It is a flexible alternative to subclassing. If it is necessary to add or withdraw responsibilities to nodes or links the decoration of an object is modified instead of the object itself. Hence, the decorator design is more flexible than static inheritance.
3. *Observer*. The observer design pattern allows one to define a one-to-many relationship between objects. If objects change their state, all the respective dependents are notified and automatically updated. This concept is applied for

[6] Further empirical research is required in order to study the distribution of benefits for different software products. This is in particular necessary with respect to the perceived derivative benefits of network effects. But such investigations are beyond the defined scope of this book. Please confer 18.1 for a discussion of this issue.

[7] Please confer the discussion on program extensions in the next subsection.

the implementation of the console, as otherwise the network visualization and the data export can decrease otherwise the performance of the simulation.

After the selection of the relevant design patterns, the program ist implemented.

11.4 Object-Oriented Implementation

The final phase is the object-oriented implementation in which the design specifications are transformed into software code. The goal is to develop a program with an optimal performance that provides the previously defined features based on the selected design patterns. In this subsection the focus is on the implementation of the customer adoption rule, i.e. the decentralized standardization rule for complex customer networks in software markets. This is relevant at this stage of the book for upcoming investigations that required to understand the mechanics of the selected customer adoption rule. But before the main routines of the simulator are presented in detail, we provide a short overview on the hardware and software infrastructure that was used for its implementation.

Hardware Infrastructure

The simulator was implemented with the following technical infrastructure.[8]

1. *Apple MacBook Pro.* Most analyses were conducted with an Apple MacBook Pro. This computer has an Intel Core Duo CPU T7500 with 2,2 GHz and 2 GB RAM.
2. *Pentium IV Compute Server.* Additional simulation data was calculated on the Pentium IV Compute Server of the Chair for Theoretical Computer Science in Mannheim. This computer is equipped with an Intel Pentium IV CPU with 3,4 GHz, an Intel Mainboard D865PERLX with 4 GB RAM PC400 DDR2 and a Seagate harddisk Barracuda ST3320613AS 320 GB SATA.
3. *Quadcore i7 Compute Server.* Large-scale complex networks were investigated on the Quadcore PC of the Chair for Theoretical Computer Science in Mannheim. This high performance computer consists of an Intel CPU Core i7-920, an Asus Mainboard Rampage II Extreme with 6 GB RAM 1066 MHz DDR3, and a VelociRaptor harddisk WD150GB SATA.

Software Infrastructure

The program was implemented in Java 1.6.0. and developed with Eclipse SDK Version 3.3.1.1. The following approved JAVA packages were used in order to implement the simulator efficiently.

[8] Many thanks to Prof. Dr. Matthias Krause and Dirk Stegemann for their support.

1. *JUNG 2.0.* The Java Universal Network Graph Framework [JUNG] is a software library for the modeling, analysis, and visualization of data which can be represented as a graph or network.[9] As it is written in Java, JUNG-based applications can benefit from the extensive built-in capabilities of the Java API, as well as other existing third-party Java libraries. The JUNG architecture is designed to support a variety of representations of entities and their relations. Moreover, it provides a mechanism for annotating graphs, entities, and relations, which make it useful in the creation of analytic tools for networks. Distribution 2.0 comprises of a number of algorithms from graph theory, data mining, and social network analysis. The library provides a visualization framework which allows to construct tools for the interactive exploration of network data.
2. *Apache Jakarta Commons Collections 3.1.* This packages is a collection of implementations, enhancements and utilities which complement the Java Collections Framework.[10] The library adds new interfaces, new implementations and utility classes which are useful in the object-oriented design of networks.
3. *Cern Colt Scientific Library 1.2.0.* This distribution comprises several free Java libraries bundled under one single uniform umbrella, such as Namely the Colt library, the Jet library, the CoreJava library, and the Concurrent library which allow to perform matrix operations, pseudo-random number generation and statistical analysis.[11] The Colt library allows the generation of data structures optimized for numerical data, such as resizable arrays, dense and sparse matrices, linear algebra, associative containers and buffer management. In addition, the Jet library contains mathematical and statistical tools for data analysis, powerful histogram functionality, Random Number Generators and Distributions useful for simulations. The CoreJava library complements the other libraries with C-like print formatting and the Concurrent library contains standardized, efficient utility classes commonly encountered in parallel and concurrent programming.
4. *Xerces Java Parser 1.4.4.* This library is used for paring XML in order to implement the GraphM for reading and writing.[12] It supports the XML 1.0 recommendation and contains advanced parser functionalities.

After this overview on the hard- and software infrastructure, the implementation of the main algorithms are described in the following subsections.

11.4.1 Network Topology

In the following, we assume a network with n actors, which we model as a directed Graph $G = (V, E)$ with a node set V corresponding to the set of actors and an

[9] For further information please confer the project website: http://jung.sourceforge.net/index.html

[10] For further information please confer the project website: http://commons.apache.org/collections/apidocs/index.html

[11] For further information please confer the project website: http://dsd.lbl.gov/~hoschek/colt/index.html

[12] For further information please confer the project website: http://xerces.apache.org/xerces-j/

edge set $E \subseteq V \times V$, where $(i, j) \in E$ if and only if actor j is influenced by actor i. Moreover, we denote by $\text{in}(j) := \{i \in V | (i, j) \in E\}$ the set of actors that influences actor j. Conversely, we define as $\text{out}(i) := \{j \in V | (i, j) \in E\}$ the set of actors that is influenced by actor i. We assume that each actor is influenced by $n \cdot b$ randomly chosen actors, $b \in [0, 1]$, and that $(i, j) \in E$ implies $(j, i) \in E$.[13]

11.4.2 Adoption Rule

We consider a single-technology case. In order to simplify the notation, we define

$$a(i,t) := \begin{cases} 1 : \text{actor } i \text{ adopts the technology in period } t \\ 0 : \text{actor } i \text{ does not adopt the technology in period } t \end{cases}.$$

An actor adopts the technology if and only if adopting induces at least as much benefit as not adopting, i.e. if and only if the benefit surplus which is equal to the net benefit of adoption is non-negative. Moreover, an actor who has not adopted in period $t - 1$ will only consider starting to adopt in period t if at least one of his influencers has adopted in $t - 1$. The net benefit of actor i in period t obtained from adopting the technology is denoted by $\text{nu}(i, t)$. This value depends on three main parameters:

1. Setup Costs $c_S(i,t)$
2. Direct Network Effects $\text{nu}^{\text{DNE}}(i,t)$
3. Indirect Network Effects $\text{nu}^{\text{INE}}(i,t)$

More precisely, we have

$$\text{nu}(i,t) = -c_S(i,t) + \text{nu}^{\text{DNE}}(i,t) + \text{nu}^{\text{INE}}(i,t)$$

and obtain the adoption rule

$$a(i,t) := \begin{cases} 1 : \text{nu}(i,t) \geq 0 \text{ and } \left(a(i,t-1) = 1 \text{ or } \sum_{j \in \text{in}(i)} a(j,t-1) > 0\right) \\ 0 : \text{otherwise} \end{cases}.$$

Please note that $a(i,\cdot)$ may not be monotonic. For example, we may have $a(i,t) = 1$, $a(i,t') = 0$, and $a(i,t'') = 1$ for $t < t' < t''$.

[13] Motivation: If i is linked to j, then j is also linked to i, but i may benefit from a confirmed contact to actor j more than j benefits from his contact to i, e.g., i = Joe Sixpack and j = Barack Obama.

11.4.2.1 Setup Costs

Adopting in period t after not having adopted in $t-1$ induces constant setup costs c_S. These have to be considered as sunk costs in periods $t' > t$, which implies

$$c_S(i,t) = \begin{cases} c_S & \text{if } a(i,t-1) = 0 \\ 0 & \text{if } a(i,t-1) = 1 \end{cases}.$$

11.4.2.2 Direct Network Effects

If actor i's neighbor j also adopts the technology in period t, direct network effects imply a benefit $u_{ij}^D \geq 0$ for i with associated costs c_{ij}^D. Hence, we have

$$\text{nu}^{\text{DNE}}(i,t) = \sum_{j \in \text{in}(i)} \underbrace{\left(u_{ij}^D - c_{ij}^D\right)}_{=:\text{nu}_{ij}^D} \cdot a(j,t)$$

with the sum taken over all of i's influencers.[14]

When making his decision for period t, i generally does not know whether his neighbor j will adopt in t, i.e., the value of $a(j,t)$ is unknown. Therefore, i estimates this value by $E[a(j,t)] \in [0,1]$, which can be interpreted as j's adoption probability for period t.

Following the decentralized standardization model of (Buxmann and König 1998; Buxmann et al. 1999; Weitzel et al. 2002), we assume[15]

$$\begin{aligned} E[a(j,t)] &:= \frac{\text{nu}_{ji}^D \cdot n \cdot b - c_S(j,t) + \text{nu}^{\text{INE}}(j,t)}{\text{nu}_{ji}^D \cdot n \cdot b} \\ &= 1 - \frac{c_S - \text{nu}^{\text{INE}}(j,t)}{\text{nu}_{ji}^D \cdot n \cdot b}. \end{aligned}$$

11.4.2.3 Indirect Network Effects

We quantify the benefit that actor i obtains from indirect network effects by $u^I(i,t) := u_i^I \cdot B(t)$ and the corresponding costs by $c^I(i,t) := c_i^I \cdot B(t)$, where $B(t) = \sum_{j \in V} a(j,t)$ denotes the number of all adopting actors in period t. We obtain

---

[14] Possible extensions are time-dependent benefits $u_{ij}^D(t)$ and costs $c_{ij}^D(t)$.

[15] Please note that $E[a(j,t)] \in [0,1]$, e.g., by forcing to one if greater than one and forcing to zero if less than zero.

$$\mathrm{nu}^{\mathrm{INE}}(i,t) = \mathrm{nu}^{\mathrm{INE}}(t) = u^I(i,t) - c^I(i,t) = \underbrace{(u_i^I - c_i^I)}_{=:\mathrm{nu}_i^I} \cdot B(t) \ .$$

Obviously, actor i needs to estimate the value of $B(t)$ by a value $E[B(t)]$. As a first approach, we assume $E[B(t)] := B(t-1)$ for $t > 0$.

11.4.3 Simplifications in the First Step

In the first step of our analysis, we assume the following simplifications.

- The number of influencers of each actor, i.e. the connectivity of the network, is constant for all network sizes. We denote this value by the constant con.
- The benefits u_{ij}^D and u_i^I as well as the costs c_{ij}^D and c_i^I are assumed to be normally distributed. Hence, the utilities nu_{ij}^D and nu_i^I are normally distributed, with parameters (μ^D, σ^D) and (μ^I, σ^I).

Under these assumptions, the expected net utility in period t for actor i is

$$\begin{aligned} E[\mathrm{nu}(i,t)] &= -c_S(i,t) + \sum_{j \in \mathrm{in}(i)} \left(\mathrm{nu}_{ij}^D \left(1 - \frac{c_S(j,t) - \mathrm{nu}^{\mathrm{INE}}(j,t)}{\mathrm{nu}_{ji}^D \cdot \mathrm{con}} \right) \right) \\ &\quad + \mathrm{nu}^{\mathrm{INE}}(i,t) \\ &= -c_S(i,t) + \sum_{j \in \mathrm{in}(i)} \mathrm{nu}_{ij}^D - \frac{1}{\mathrm{con}} \sum_{j \in \mathrm{in}(i)} \left(\frac{c_S(j,t) - \mathrm{nu}^{\mathrm{INE}}(j,t)}{\mathrm{nu}_{ji}^D} \right) \\ &\quad + \mathrm{nu}^{\mathrm{INE}}(i,t) \end{aligned}$$

11.4.4 Package Structure of the Implementation

11.4.4.1 Package *network*

The package *network* contains the classes and interfaces for representing a network (please confer Fig. 11.2). A network consists of actors and links between the actors, which are represented by the classes *Network*, *Actor*, and *Link*. An *Actor* a may act as an influencer of *Actor* b (indicated by a directed *Link* from a to b). In this case, a is b's influence. Based on an *AdoptionRule*, an *Actor* decides whether or not to adopt a certain *Technology*.

Classes implementing the *NetworkGenerator* interface produce *Networks* with particular properties. *PowerLawNetworkGenerator*, *RandomNetworkGenerator* and

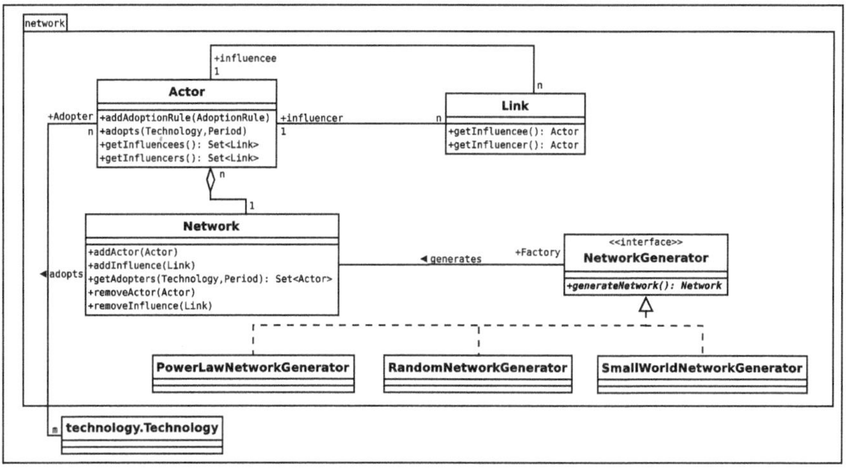

Fig. 11.2 UML diagram of the package network
Source: Author

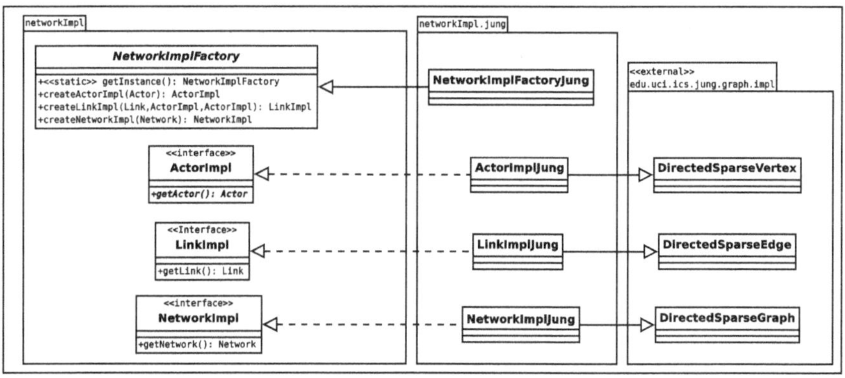

Fig. 11.3 UML diagram of the packages network.impl and network.impl.jung
Source: Author

SmallWorldNetworkGenerator compute the types of networks that are most relevant for our application.

11.4.4.2 Packages *network.impl* and *network.impl.jung*

Figure 11.3 gives an overview of the packages *network.impl* and *network.impl.jung*.

In order to maximize an application's performance, it is often reasonable to use highly optimized third-party libraries for special purposes. Therefore, the package *network.impl* defines an adaptor layer that allows to use arbitrary third-party network implementations with the generic interface of the package *network*. Partic-

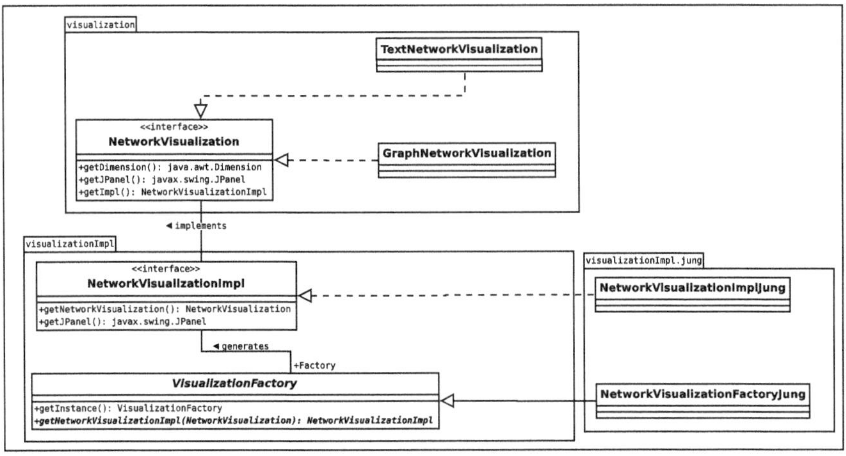

Fig. 11.4 UML diagram of the packages visualization, visualization.impl and visualization. impl.jung
Source: Author

ularly, the package *network.impl.jung* provides adaptor classes for the Jung network library. Which library is actually used when implementations are requested, is determined in the static method *getInstance()* of the class *NetworkImplFactory*, which returns the *NetworkImplFactory* for the respective library.

11.4.4.3 Packages *visualization*, *visualization.impl* and *visualization.impl.jung*

In order to get an intuitive impression of the diffusion process, a graphical representation of the network and the adopting actors is needed. This functionality is provided by the package *visualization* and the corresponding implementation packages *visualization.impl* and *visualization.impl.jung* (Please confer Fig. 11.4).

For network visualizations even more than for networks itself, it is reasonable to rely on third-party libraries. Therefore, the *visualization* package abstracts from particular implementations and adopts them by concrete instances of the adopter classes in the package *visualization.impl*, for instance by the package *visualization.impl.jung* that allows to use the visualization infrastructure of the Jung network library.

11.4.5 Package network.statistics

Networks can be characterized by various parameters. In our implementation, these parameters are represented as subclasses of the abstract class *NetworkParameter*. In the current implementation, *MeanPathLength*, *AverageDegree*, *ClusteringCoefficient*, *GiantComponent* and *Diameter* are supported (please confer Fig. 11.5).

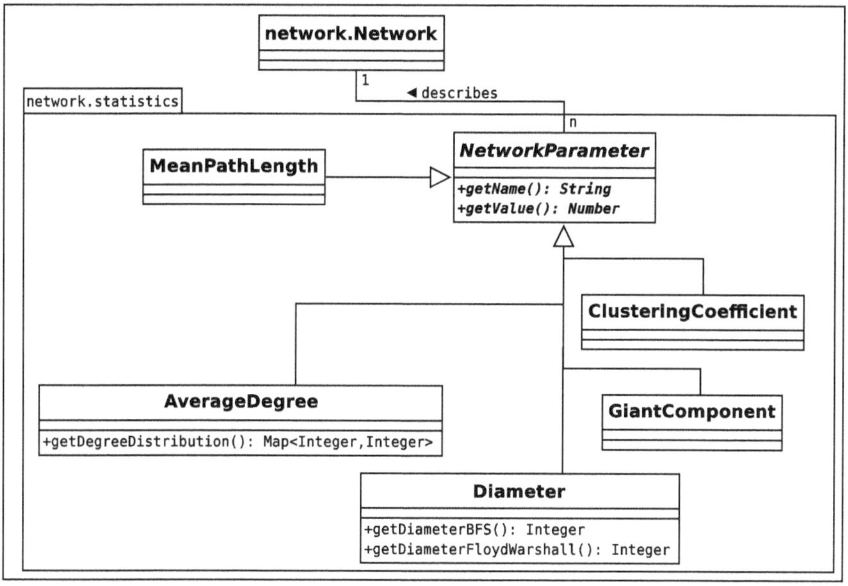

Fig. 11.5 UML diagram of the package network.statistics
Source: Author

11.4.6 Package technology

The package *technology* provides the main classes for the simulation of the diffusion process (Please confer Fig. 11.6). In order to make an adoption decision for a particular *Technology* in a certain *Period*, an *Actor* needs to apply an *AdoptionRule*. For performance reasons, a *CachingAdoptionRule* remembers the adoption decisions of earlier *Period*s and avoids recomputation. Particularly, *ThresholdAdoptionRule* implements a simple adoption rule based on the number of adopting influencers, and *BasicAdoptionRule* encodes the adoption rule described in Section 11.4.2 with constant costs and utilities, while *BasicAdoptionRuleDistr* implements the same rule with normally distributed cost and benefit parameters. To allow for flexible extensions of the model, the parameters of the adoption rule are implemented as decorators of *BasicAdoptionRuleDistr*.

The simulation of a single diffusion process is carried out by instances of the class *DiffusionSimulator*, which triggers the computation of the actors' adoption decisions for all *Period*s in a specified interval under fixed parameter values. *MultipleDiffusionSimulator* computes simulations for varying parameter values by performing a *DiffusionSimulator*-based simulation for each parameter combination. The concrete *MultipleDiffusionSimulator* subclasses *SimpleDiffusionSimulator* and *SingleNetworkDiffusionSimulator* implement the particular multiple diffusion process of our application.

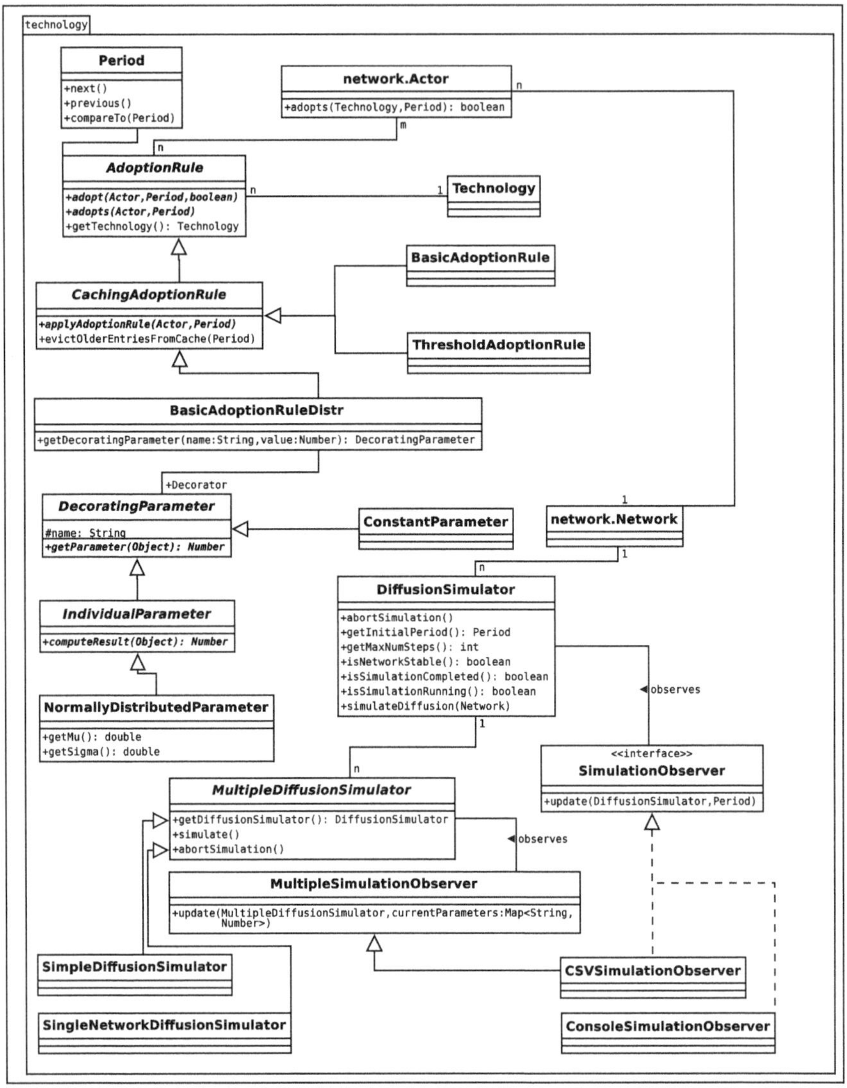

Fig. 11.6 UML Diagram of the Package technology
Source: Author

DiffusionSimulators and *MultipleDiffusionSimulators* provide information about their status and intermediate simulation results to classes implementing the *SimulationObserver* and *MultipleSimulationObserver* interfaces. Our application writes simulation results to CSV-Files and the console through the classes *CSVSimulationObserver* and *ConsoleSimulationObserver*, which both implement this interface.

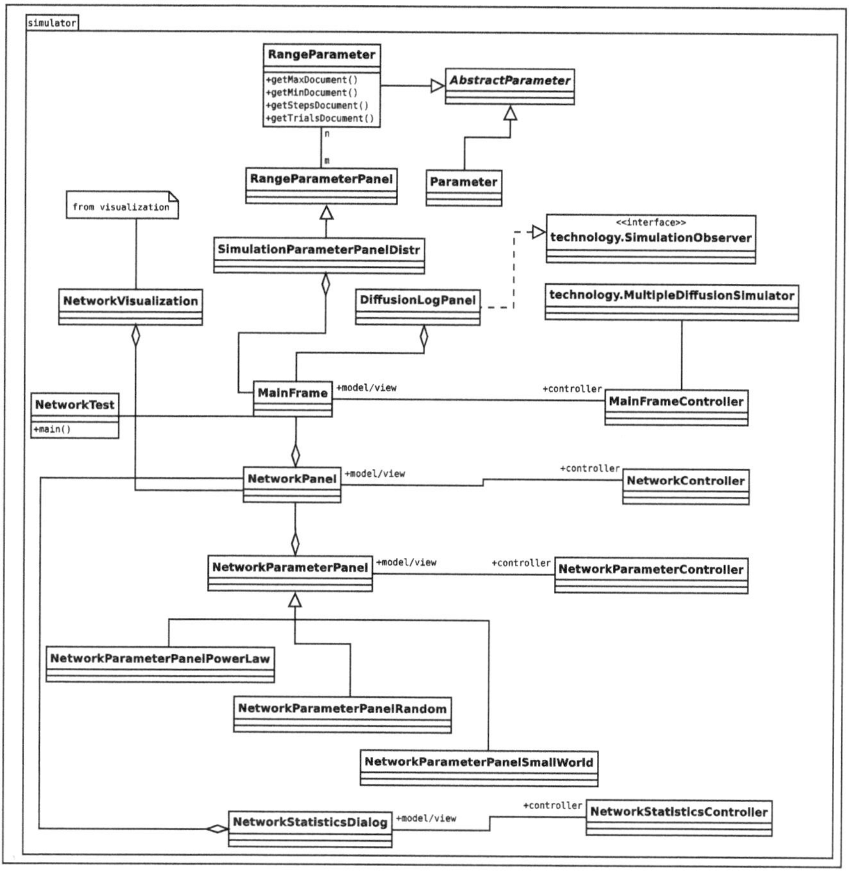

Fig. 11.7 UML diagram of the package simulator
Source: Author

11.4.7 Package simulator

The package *simulator* defines the application's graphical user interface and relies heavily on the model-view-controller (MVC) pattern (please confer Fig. 11.7).

The application's *MainFrame* consists of the *NetworkPanel* with a *NetworkParameterPanel* for the various network generation algorithms and the *NetworkVisualization*, as well as the *SimulationParameterPanel*. Additionally, it contains an area displaying various logging information, the *DiffusionLogPanel*. The *RangeParameterPanels* allow to specify a minimum and a maximum value, a step width and the number of trials for each *Parameter*. The *NetworkParameter* classes from the package *network.statistics* display their parameter values in the *NetworkStatisticsDialog* upon the user's request. Finally, the package specifies the class *NetworkTest*, which contains the application's *main* method.

11.5 Reconsideration of the Simulator

The presented complex networks adoption and diffusion simulator is a powerful tool for network theoretical investigations. It was designed according to the object-oriented programming paradigm based on approved high-performance libraries, driven by computational requirements to optimize the performance of the applied algorithms. Thereby, the program provides a high level of flexibility with respect to further modifications of individual components and a considerable computational performance. Due to its flexibility, it can be applied for various purposes. For example, it is used in the following section to investigate the network theoretical hypotheses relevant to the open research question on customer network-centric valuation in software markets. But it can also be used more general, to analyze properties and dynamics in complex customer networks even under consideration of varying network topologies. Therefore, it is a cornerstone of the complex networks framework for valuation in software markets that is designed in one of the following chapters. But despite of the broad spectrum of functionalities, the adoption and diffusion simulator is limited by the following most relevant constraints.

1. *Network Size*. Programming requires one to handle a trade-off between performance and extendibility. The simulator is designed with a general network definition according to the outlined abstract factory design pattern in order to allow flexible adjustments and extensions with additional libraries, packages and classes if required. This decision implies that the network computation and representation is not optimized for performance. Thus, the maximum network size of the simulator is limited and depends on the simulated network topology and on the computational performance of the computer. For example, it is important to note that the required computational performance of random networks vitally depends on the chosen connectivity. While it is possible to create very large but relatively sparse networks, denser networks have to be smaller.[16]
2. *Customer Adoption Rule*. The outlined decentralized standardization model is implemented although other rules are also plausible. But as the rule is a capsulated module of the simulator it is possible to exchange it against other suitable concepts, e.g. a mean-field approximation. For this reason, research should implement further adoption rules. If multiple adoption rules are available it is possible to compare their explanatory potential based on real world observations.

Despite such constraints, the adoption and diffusion simulator remains a suitable tool for a complex networks analysis. Please note that the depicted limitations can be resolved based on additional research efforts, which are, however, beyond the scope of this book. With respect to the limited network size it important to note that the available computational power increased rapidly over the last decades and

[16] Under the given computational restrictions, the simulator allowed one to simulate a random network of 300,000 nodes with an average connectivity of 5. A network with an average connectivity of 50, in turn, is limited to a network with 50,000 nodes.

there are significant reasons to believe that this trend will continue on a short- and medium-term time horizon.[17] Therefore, it is reasonable to assume that it is a matter of time until more computational power and memory are available that will allow to overcome the existing limitations for simulating very large-scale or even real-size models.[18] As far as the limitations of the customer adoption rule are concerned, it is possible to conduct further modifications due to the modular object-oriented design. Accordingly, it is possible to incorporate further network topologies, further customer adoption rules, further visualization possibilities, e.g. visualization packages for large-scale networks, or further network theoretical measures. After this reconsideration of the simulator, it is used in the following chapter for investigations of the previously developed hypotheses in the complex networks analysis of complex customer networks.

[17] The computational power of the fastest computers increased by a factor of thousand during the last 10 years (Heise 2008).

[18] Please note that today highly specific scientific projects have already access to high performance computers and high performance computing fnetworks, e.g. JUGENE in Jülich (Heise 2008).

Chapter 12
Complex Networks Analysis of Customer Networks

"The Oracle Says: Presley, Elvis has a Bacon number of 2: Presley, Elvis was in *King Creole* (1958) with Matthau, Walther Matthau, Walther was in *JFK* (1991) with Bacon, Kevin" Excerpt from the website "The Oracle of Bacon"

by Brett Tjaden and Glenn Wasson

Complex networks theory incites the formulation of the hypotheses for customer network-centric valuation in software markets. In this chapter the respective hypotheses are investigated from a complex networks perspective supported by the developed simulator. The main research areas of interest for valuation in software markets are investigated. These comprise diffusion dynamics in varying network topologies, scaling properties and network topologies of customer networks, contributions to valuation in software markets, and limitations due to the social nature of customer networks. These research topics are pursued in the following sections.

12.1 Diffusion Dynamics in Varying Network Topologies

In this section variations of network topologies are investigated. For this purpose, the following three steps are conducted. First, the research hypothesis is formulated and its motivation is depicted. In a second step, the hypothesis is challenged in an analysis based on complementary theoretical and numerical research. Finally, the most important findings are summarized in the reconsideration of the hypothesis.

12.1.1 Hypothesis on Diffusion Dynamics In Varying Network Topologies

Complex networks research identified a variety of network topologies (Watts 1999; Albert 2001; Barabasi et al. 2000). As such topologies have an impact on the diffusion behavior of other networks, a fundamental research question is whether this is also relevant for customer networks in software markets. Based on the existing incidences, we formulate the hypothesis that the variation of customer

A. Kemper, *Valuation of Network Effects in Software Markets*, Contributions to Management Science, DOI 10.1007/978-3-7908-2367-7_12,

network topologies has an impact on the adoption and diffusion behavior in software markets [HP1]. The motivation to formulate this hypothesis is to link the previous investigations on valuation, network effects and complex networks. It the network topology is relevant to developments of customer networks, they would also have an impact on sales, revenues, and cash flows, which, in turn, would influence valuation in software markets.

12.1.2 Analysis of Diffusion Dynamics In Varying Network Topologies

While there are investigations on network topologies in various research areas, we want to prove their relevance specifically for customer networks in software markets by contraposition. Hence, it is necessary to prove that the diffusion dynamics of customer network vary with the underlying network topologies. This would imply that diffusion dynamics in varying network topologies are not equal. A popular approach to investigate the relevance of network topologies is to demonstrate that their metrics are different. If network topologies have different network properties and if the properties influence the diffusion dynamics, it is reasonable to conclude that different network topologies have an impact on the diffusion dynamics. Accordingly, relevant network properties of small-world networks, scale-free networks, and random networks are compared.[1] After these investigations of relevant structural network properties, we shift the focus of the analysis to the impact on diffusion dynamics in software markets.[2] The comparison of the network topologies is conducted in two steps. First, small-world networks are compared to random networks, before the random graphs are compared to scale-free networks.

Network Properties of Small-World Networks

A Small-World Network Analysis between small-world graphs and random networks reveals that there are considerable differences in their network metrics. The most relevant aspects of a theoretical and numerical Small-World Network Analysis are depicted in the following.[3]

[1] Please note that random networks are frequently used in order to benchmark the behavior of other networks. A main reason for this choice is that random networks are assumed by default many traditional approaches, e.g., analytical approaches, which ignore the internal structure of networks. This implicit assumption is equivalent to assuming homogeneous random networks.

[2] Please note that the initial investigations on the properties of network topologies are required. They help to grasp the differences between the network classes and reveal the importance of network metrics that are used throughout the book.

[3] Please confer Sect. 10.3.2 for further details on the Small-World Network Analysis.

1. Small-World Network Analysis of Small-World Graphs

In a Small-World Network Analysis the development of the clustering coefficient and of the characteristic path length of networks are compared for varying rewiring probabilities. The characteristic path length, i.e., the mean average path length, is relevant to diffusion dynamics as it measures the required distances that a diffusion has to traverse in order to reach each node in a network. Accordingly, a lower characteristic path length of a network implies, ceteris paribus, a faster diffusion process. Research on small-world networks reveals that they tend to change their behavior.[4] They undergo a transition from a large-world phase, in which the average distance scales linearly with its size, to a small-world phase, in which the distance decreases logarithmically. Random networks, in turn, are not effected by random rewiring as they are generated by random wiring.[5] All in all, the theoretical deductions confirm the hypothesis that diffusion dynamics of small-world networks and random networks are different. In the following section the theoretical reasoning is supported by the simulations and analyses that illustrate the mechanisms and dynamics of networks.

2. Numerical Small-World Network Analysis of Small-World Networks

In addition, to the previously derived theoretical contributions, our comparison of small-world networks and random networks support the findings of the previous analysis. Accordingly, the development of the characteristic path length and of the respective clustering coefficient are investigated for varying rewiring probabilities in both network topologies, before the results are compared. All of the following experiments are conducted with the previously developed complex networks adoption and diffusion simulator. For this purpose, we create a standard network with the following characteristics. The size of the network is set to 1,000, its connectivity to 10, and the default rewiring probability is assumed to be 0 for random and scale-free networks. Small-world networks are rewired by default with a rewiring probability of $\beta = 0.5$ unless it is otherwise stated.

The *Small-World Network Analysis of the Small-World Network* shows that the relevant network metrics develop for a varying rewiring probability as follows. The characteristic phase length reveals a sudden and rapid phase transition very close to the beginning of the rewiring process even for very small β.[6] For a more detailed overview on the exact behavior of the characteristic path length in small-world graphs, it is necessary to plot its development on a logarithmic scale for varying rewiring probabilities.[7] The clustering coefficient, in turn, remains high, even long

[4] Please confer (Watts 1999).

[5] Please confer Sect. 10.3.1.

[6] Please confer Fig. 12.1.

[7] Please confer Fig. 12.2.

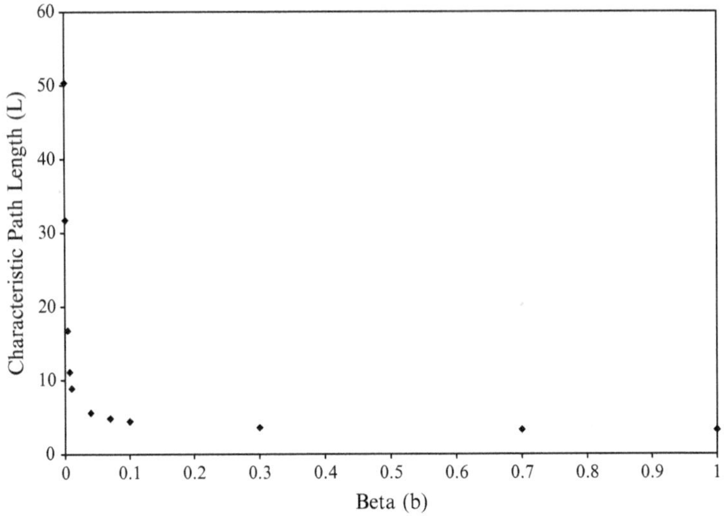

Fig. 12.1 Small-world network analysis: Linear characteristic path length of SWN for varying rewiring probabilities
Source: Author

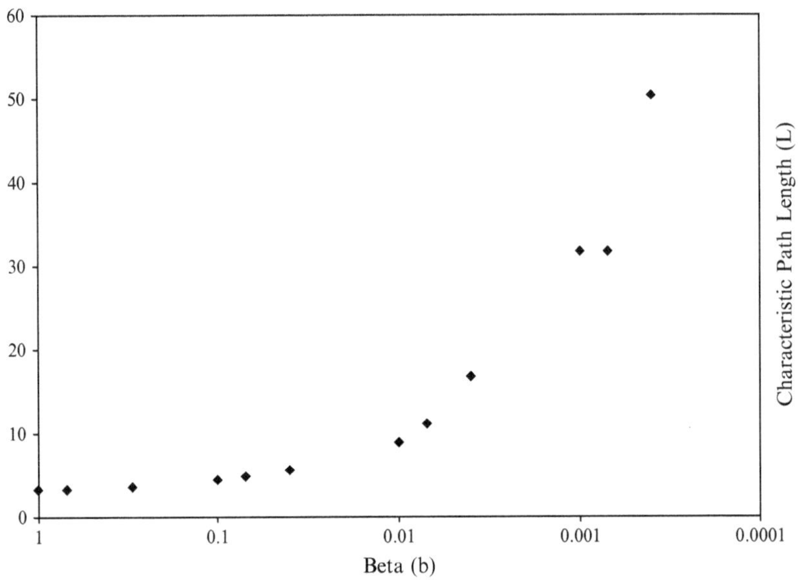

Fig. 12.2 Small-world network analysis: Logarithmic characteristic path length of SWN for varying rewiring probabilities
Source: Author

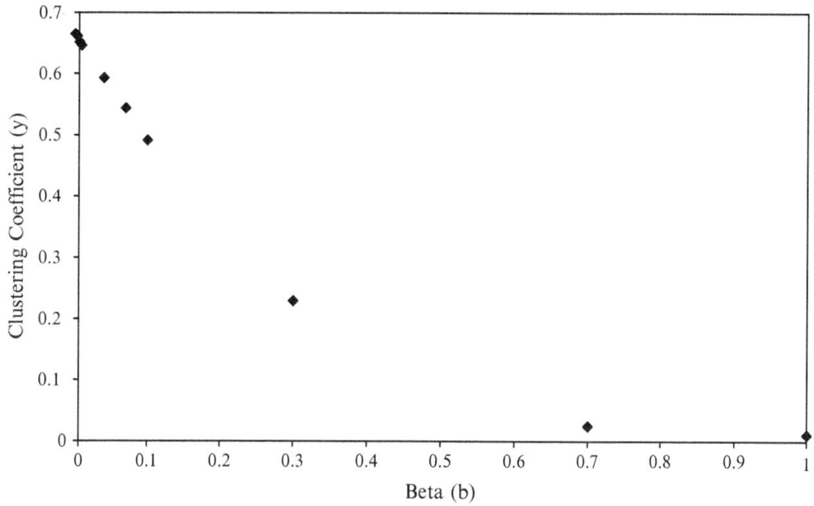

Fig. 12.3 Small-world network analysis: Linear clustering coefficient of SWN for varying rewiring probabilities
Source: Author

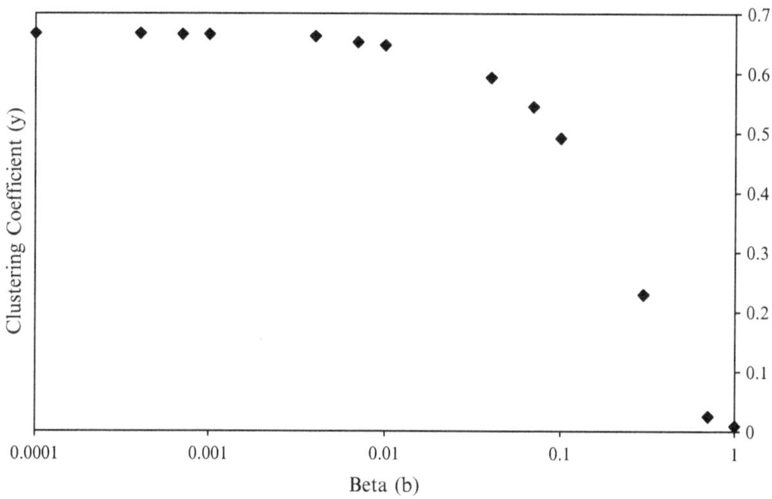

Fig. 12.4 Small-world network analysis: Logarithmic clustering coefficient of SWN for varying rewiring probabilities
Source: Author

after the characteristic path length has already approached an asymptote (Figs. 12.3 and 12.4).

Details on the rapid development of the clustering coefficient for low β can be observed in a logarithmic representation. A combination of both observations reveals the typical development of the characteristic path lengths and of the clustering

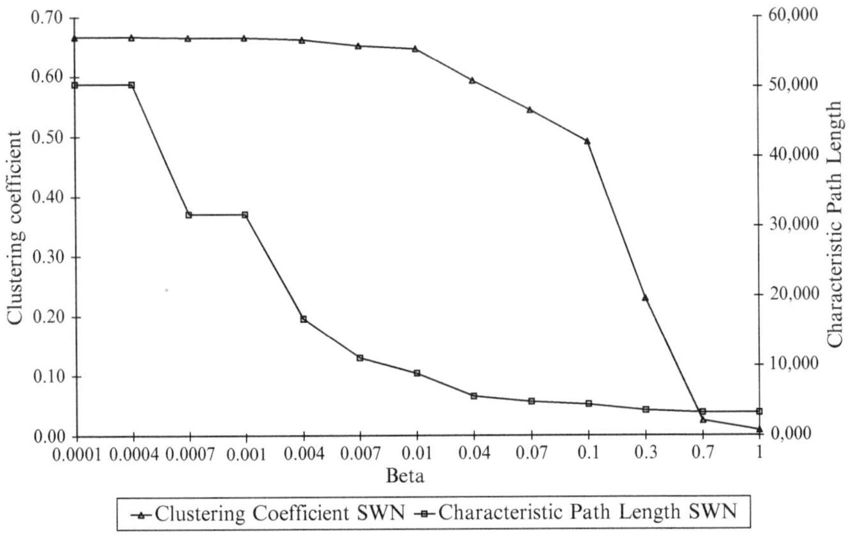

Fig. 12.5 Small-world network analysis: Characteristic path length vs. clustering coefficient for varying rewiring probabilities in small-world networks
Source: Author

coefficient in a small-world network.[8] The analysis reveals that the characteristic path length decreases rapidly if a few short-cuts are added to the graph. The clustering coefficient, in turn, remains high and converges finally for large β to an asymptote. Although this behavior has to occur at some stage, the intensity of this reaction to rewiring is a surprising phenomenon. Please note that this phase transition is purely topological in nature. The number of links is conserved and only their position is changed during the rewiring process.

A numerical *Small-World Network Analysis of a random graph* which is created with similar parameters reveals a different profile. Accordingly, the characteristic path length and the clustering coefficient of the random network are constant for all rewiring probabilities β. This finding is not a surprise as random networks are constructed by random wiring of nodes (Fig. 12.6).

Finally, the results of the independently pursued Small-World Network Analyses of Small-World Networks and of Random Networks are integrated. This *comparison* is presented in Fig. 12.7. The analysis exhibits that the clustering coefficient of the small-world network remains high, even long after the characteristic path length has already approached the random graph asymptote. This development reveals that the characteristic path length of small-world networks interpolates between a ring lattice for small β and a random network for β close to 1. Hence, small-world networks reveal topological characteristics in the intermediate regime between a ring lattice and a random network. These findings prove small-world networks and random networks are not identical with respect to all parameters.

[8] Please confer Fig. 12.5.

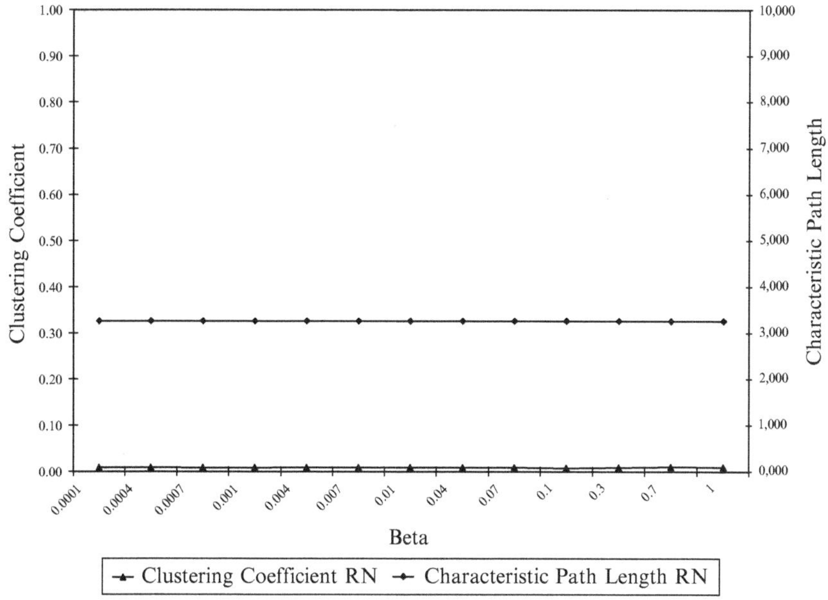

Fig. 12.6 Small-world network analysis: Characteristic path length vs. cluster coefficient for varying rewiring probabilities in random networks
Source: Author

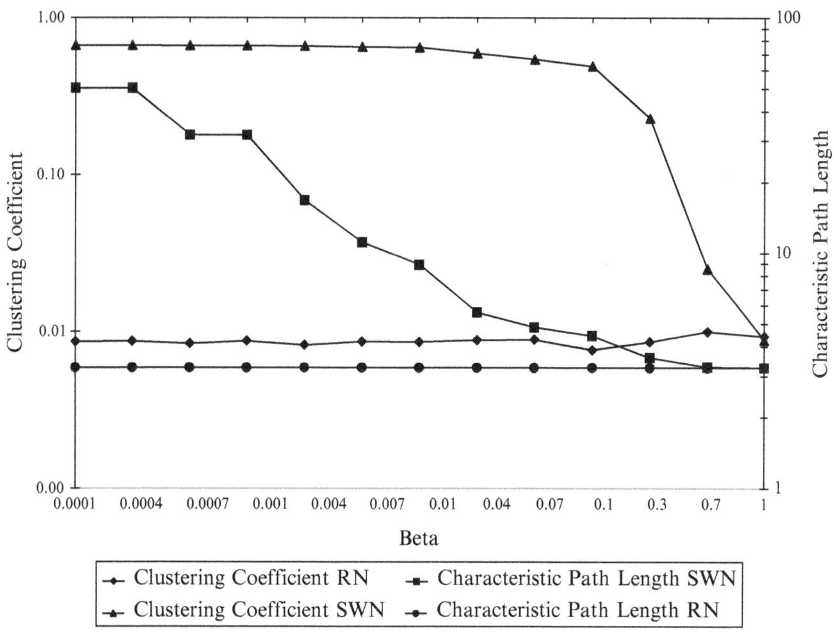

Fig. 12.7 Small-world network analysis: Comparison of small-world analyses of RN and SWN
Source: Author

3. Summary of Findings for Small-World Networks

All in all, our Small-World Network Investigations reveal significant differences between small-world networks and random graphs. These differences result from a different behavior of central network metrics. A central observation is that the clustering coefficient of the small-world network remains high, even long after the characteristic path length has already approached the asymptote of the random graph. This is the classical small-world network phenomenon. In addition, the investigations reveal that small-world networks interpolate from a topological perspective between a ring lattice for small β close to 0 and a random network for β close to 1. While small-world networks adopt the topological characteristics of random networks for $\beta = 1$, both topologies exhibit different network properties for all other rewiring probabilities. These findings are also confirmed by studies on SIS-models or SIR-models in cellula automata (Boccara and Fuks 2003).

Despite of these insights, further investigations are required in order to assess the implications of these findings for diffusions in customer networks in software markets. Such an analysis is pursued in an upcoming section after a comparative Small-World Network Analysis of scale-free networks and random networks.

Network Properties of Scale-Free Networks

Similar to the previous research on small-world networks, we compare the relevant network metrics of scale-free networks and random networks in a Small-World Network Analysis.[9] The purpose of this analysis is to compare, again, relevant network properties that may support the hypothesis that diffusion dynamics are different in varying network topologies.[10] First, some theoretical findings are considered, based on which the respective numerical simulations are conducted. A comparison of dynamics in both network topologies concludes this section.

1. Propagation and Non-Zero Percolation in Scale-Free Networks

According to an influential school of thought in complex networks research, diffusions in scale-free networks always propagate (May and Anderson 1988; Pastor-Satorras and Vespignani 2001b; Boguna et al. 2002; Pastor-Satorras and Vespignani

[9] The investigations are conducted with the adoption and diffusion simulator. Please note that we conduct the following simulations with the following standard network parameters. The size of the network is set to 1,000 its connectivity to 10. Small-world networks are rewired by default with a rewiring probability of $\beta = 0.5$ unless this is otherwise stated. The number of links for scale-free networks is set to 500 in order to generate a scale-free network with the same connectivity, i.e., 10, than the other topologies.

[10] Please confer Sect. 10.3.2 and the previous section for further details on the Small-World Network Analysis.

2003; Newman 2003b). Research suggested that *propagation* in scale-free networks occurs regardless of transmission probabilities between individuals, if the exponent of the powerlaw degree distribution is less than 3.[11] As most scale-free networks satisfy this condition, it is reasonable to assume universal propagation characteristics of scale-free networks. This insight was first derived by research on computer virus epidemiology, but other research on epidemics revealed similar insights in various research areas (May and Anderson 1988; Pastor-Satorras and Vespignani 2001b; Pastor-Satorras and Vespignani 2003). In the original study, fully mixed differential equation models of epidemics were investigated. Although the network structure was not explicitly considered, the total population was clustered into strata with respect to their infection rate (May and Anderson 1988). Thereby, the study allowed one to prove that the variation of the number of infectives over time depends on the variance of this rate over the respective classes (May and Anderson 1988). Moreover, the experiments revealed that diseases always multiply exponentially, if the variance diverges, which is the case in scale-free networks with an exponent less than 3. Consequently, scale-free networks were assumed to be particularly susceptible to diffusions (Boguna et al. 2002). But this conclusion was revised based on further complex networks research Eguiluz and Klemm 2002; Blanchard et al. 2002). In particular, further complex networks studies revealed a *non-zero percolation threshold* for certain types of correlations between nodes if the network has high transitivity (Warren and Sokolov 2002; Eguiluz and Klemm 2002). In summary, this discussion reveals that previous investigations of scale-free networks discovered diffusion dynamics that differ from those of other network topologies. These theoretical derivations are supported by numerical Small-World Network Analyses that are conducted in the following chapter.

2. Numerical Small-World Network Analysis of Scale-Free Networks

The Small-World Network Analysis of scale-free networks has, again, a focus on the development of the characteristic path length and of the respective clustering coefficient for varying rewiring probabilities. For this purpose, the relevant network properties of scale-free networks are simulated and analyzed, before the results are compared to the previously derived findings on random networks.

The results of the *Small-World Network Analysis of scale-free graphs* indicate that rewiring has no impact on the investigated network characteristics. Neither the clustering coefficient nor the characteristic path length notably change for varying rewiring probabilities β.

In a second step, the results of the previously conducted *Small-World Network Analysis of random graphs* are reconsidered. Accordingly, the characteristic path length and the clustering coefficient of the random network are constant for all rewiring probabilities β. At the same time, the analysis demonstrates that the

[11] Please confer Sect. 10.3.3.

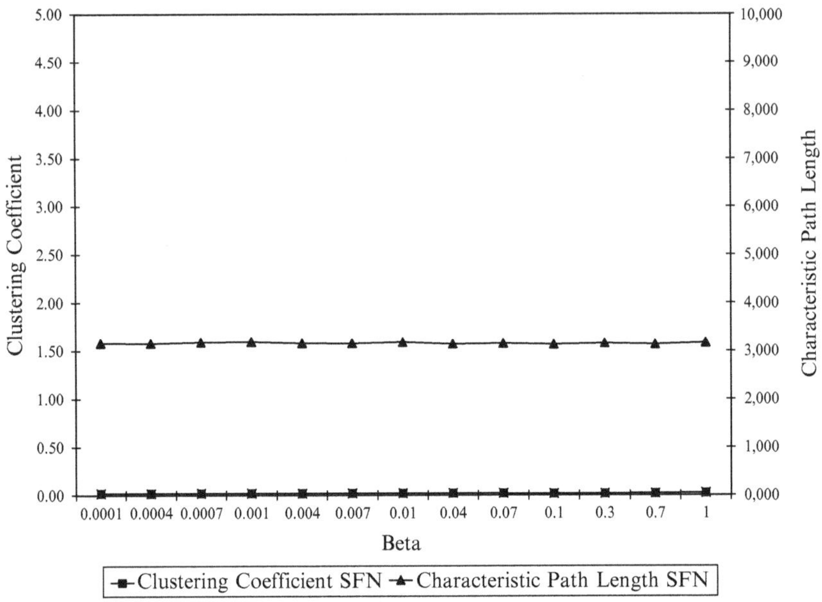

Fig. 12.8 Small-world network analysis: Characteristic path length vs. cluster coefficient for varying rewiring probabilities in scale-free networks
Source: Author

random networks are not clustered, as the respective clustering coefficient is an asymptote at a very low level.

A *comparison* of the results from both Small-World Network Analyses reveals that both topologies have a similar profile with respect relevant network metrics, i.e., the characteristic path lengths and the clustering coefficient.

It is interesting to note that the level of parallel developments is slightly different. The characteristic path length of the random network varies between 3, 257 and 3, 261, while the respective value of scale-free networks varies between 3, 134 and 3, 181. Similarly, the clustering coefficients of random networks, i.e., 0, 01, and of scale-free networks, i.e., 0, 02 are nearly similar. Since both characteristics reveal a parallel development of relevant network metrics on slightly different levels, it is reasonable to deduce that their diffusion dynamics are similar (Figs. 12.8 and 12.9).

3. Summary of Findings for Scale-Free Networks

The analysis of the diffusion dynamics in scale-free networks supports the hypothesis that there are considerable differences among the investigated topologies. The discussion of propagation and non-zero percolation thresholds in scale-free networks indicates that the properties of the network topologies can also have a decisive impact on the diffusion dynamics. Supportive numerical simulations revealed that the investigated network metrics of scale-free networks and random networks are

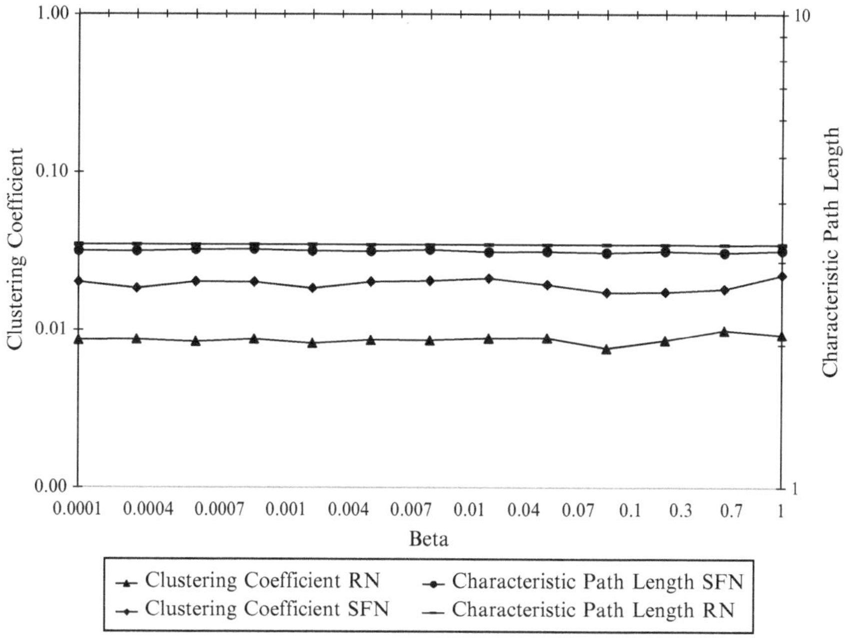

Fig. 12.9 Small-world network analysis: Comparison of small-world analyses for RN and SFN
Source: Author

similar at slightly different levels. Therefore, it is reasonable to assume that the respective diffusion dynamics in both topologies are also similar. This, however, would imply that the properties of small-world networks and scale-free networks are also very different. But in order to prove the hypothesis and its validity for software markets, further investigations are required. Therefore, additional simulations are conducted with the adoption and diffusion simulator. The focus of this research is directly on diffusion dynamics of relevant network topologies that prove our hypothesis.

Diffusion Dynamics in Software Markets

The previous Small-World Network Analyses on relevant network properties indicate that some topologies have similar characteristics, e.g., scale-free networks and random networks, while others are different, e.g., small-world network vs. random networks. Hence, varying network topologies are likely to have an impact on diffusion processes in networks. Now we extend the research approach. We transfer the initial findings on the properties of networks to customer networks in software markets based on further numerical simulations. While we investigated previously rather static network metrics, the focus of the following research is switched directly to diffusion dynamics in complex networks. The underlying motivation is to derive further insights on networks that may be helpful to enhance valuations in software markets based on a better understanding of network effects. Accordingly, we pursue a prove by contraposition. For this purpose, we simulate and analyze ceteris paribus the diffusion processes in all relevant network topologies. If the network topologies were irrelevant, the simulations would deliver ceteris paribus identical results. In the following, the investigations are structured according to the respective network topologies before the topologies are compared at a later stage after the general diffusion processes are assessed in the remaining time.

1. General Diffusion Process and Parameters

The following investigations are conducted ceteris paribus in varying network topologies. All simulations share the following underlying assumptions.

1. *Network topology.* The network topology is the only parameter that is changed between the three most relevant classes of networkss.
2. *Population.* All networks are assumed to have a population $N = 1{,}000$ with an average connectivity of 10. The population consists of homogeneous customers.[12]
3. *Adoption rule.* We assume that the previously presented decentralized standardization rule is capable to explain the most relevant factors of purchasing decisions in software markets.[13]
4. *Adoption parameter.* In the standard diffusion process of our investigations, the following parameters are assumed: Set-up costs are set to 100, the average direct utility is set to 50, the indirect utility is set to 10 and the number of initial adopters is set to 10. Initially, no values are assumed for the direct utility, direct costs, indirect costs and variation of the indirect benefit.
5. *Network size.* We have conducted investigations on all topologies for network sizes ranging from 1,000 to 5,000 with iterations steps of 1,000. In addition, we simulated networks ranging from 10,000 to 50,000 with steps of 10,000.
6. *Time horizon.* In the following, a standard time horizon of 60 iterations is assumed.

2. Diffusion Results in Random Networks

In order to investigate diffusion dynamics of random networks, we conducted five simulations with different network sizes ranging from 1,000 to 5,000. The results of the simulations are depicted for the standard time horizon of 60 periods in the following Fig. 12.10. In order to provide an long-term overview on the relevant network diffusion dynamics, we depict the results for the chosen simulation time horizon of

---

[12] Please note that the simulator provides also the possibility to implement stochastic distributions of benefits as depicted in Sect. 12.4.

[13] Please confer Sect. 7.2.2.2 for the description and Sect. 11.4.2 for a description of its algorithmic implementation.

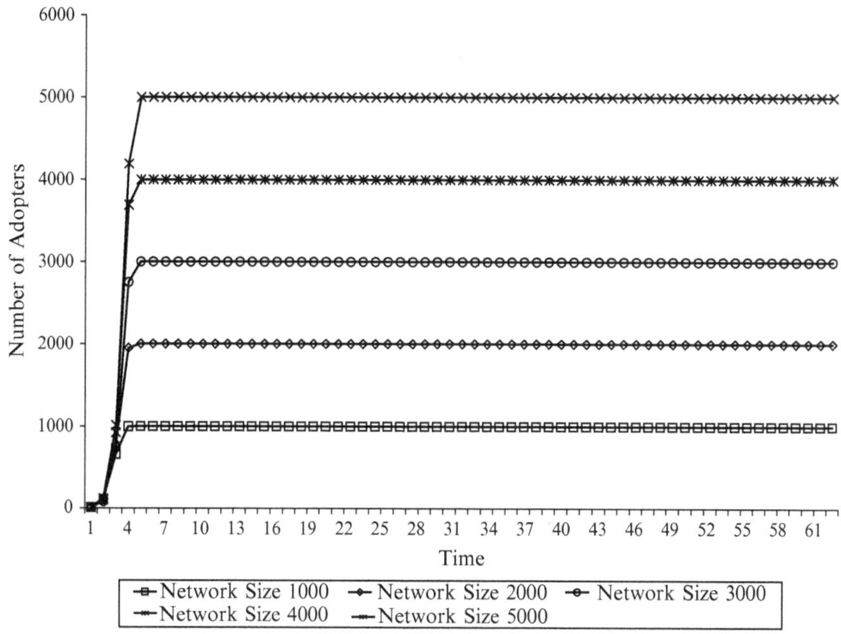

Fig. 12.10 Diffusion dynamics of random networks: Diffusions dynamics in random networks for 60 periods
Source: Author

60 periods in Fig. 12.10. This plot reveals that the diffusions increase rapidly and reach finally to total population. Given the applied parameters only a few iterations are necessary to infiltrate the whole network. After this overview, we depict the most relevant developments of the diffusion process in Fig. 12.11. This summarizes the diffusion dynamics in the simulated networks over eight periods. The shorter time period reveals a higher level of detail on the beginning of the diffusion. For a network size of 1,000 nodes, the full population is reached after four iterations, while it tool five iterations to penetrate the network at size 5,000. The graph emphasizes the explosive diffusion from iteration three to round four.

3. Diffusion Results in Scale-Free Networks

In this section, we investigate the same diffusion process on scale-free networks. The only change concerns the topology of the network. For this purpose, we simulate again five networks with a size ranging from 1,000 to 5,000 for a total time horizon of 60 periods. Figure 12.12 depicts the diffusion data. Similar to the behavior of random networks, it is possible to observe that the diffusion expands already quickly in the early phases of the diffusion process. Hence, we depict the initial phase of the diffusion process in Fig. 12.11. The plot summarizes the diffusion dynamics on the simulated networks for ten periods. The details of the initial phase reveal that

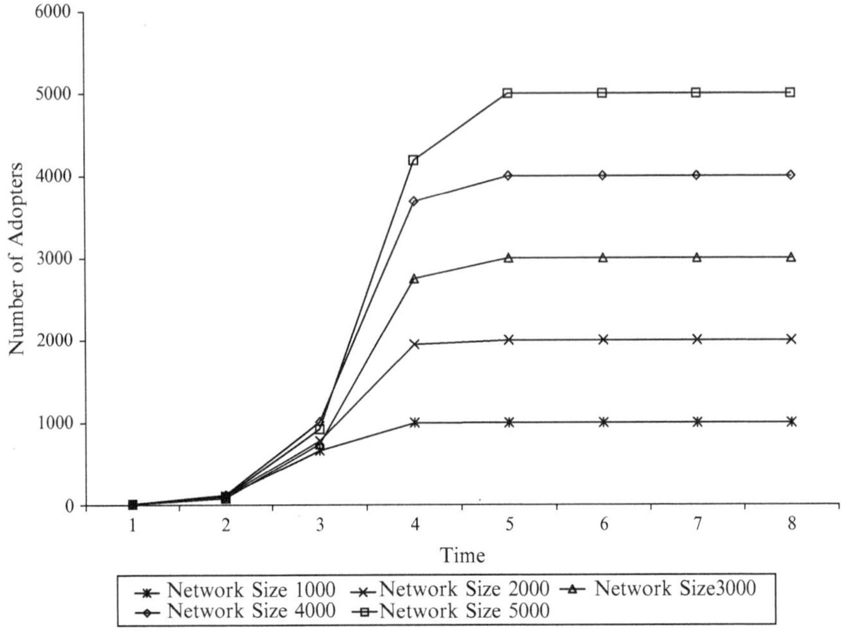

Fig. 12.11 Diffusion dynamics of random networks: Diffusions dynamics in random networks for eight periods
Source: Author

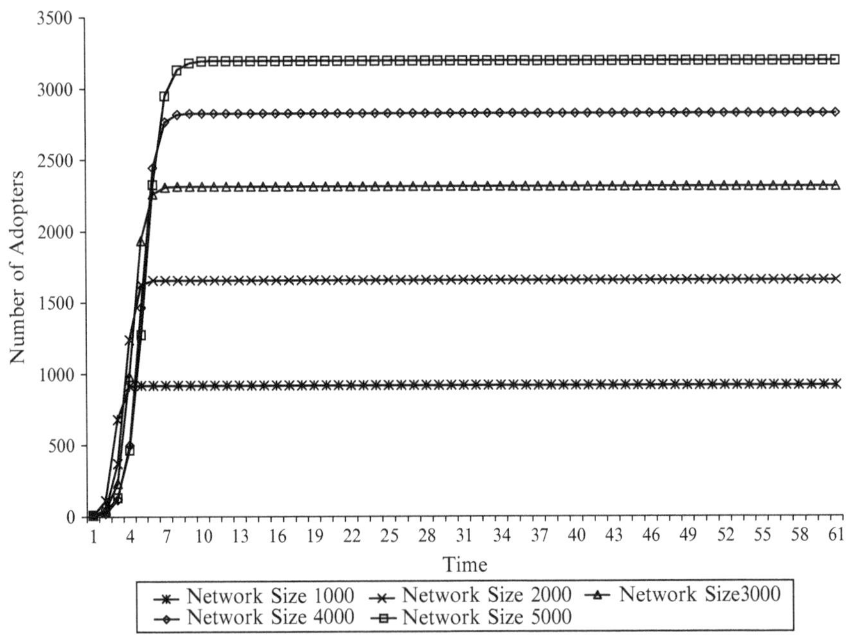

Fig. 12.12 Diffusion dynamics of scale-free networks: Diffusions dynamics in scale-free networks over 60 periods
Source: Author

the diffusion takes longer than on random networks, while network size 1,000 is reached after four iterations, the maximum network size slightly above 3,000 in the 5,000 node network is reached after nine rounds.

Moreover, we simulate diffusions in scale-free networks that have a size that ranges from 10,000 to 40,000 at a step size of 10,000. The parameters of the adoption rule, in turn, are not changed. The results of the large-scale investigations are summarized in Fig. 12.22. The plot reveals that much larger networks reveal similar dynamic diffusion dynamics than the smaller networks.

4. Diffusion Results in Small-World Networks

Finally, we investigate the same diffusion process on small-world networks. For this purpose, we simulated, again, networks with a size ranging from 1,000 to 5,000 in intervals for 1,000 over a 60 periods time horizon. The visual impression is different from the earlier observations. Although the diffusions occur under the same conditions, the process takes much longer (Figs. 12.13–12.15).

Figure 12.16 has a focus on the beginning of the diffusion which is simulated for 13 periods in the small-world networks. The graph reveals that the slopes of the growth rates are much lower than in random or scale-free networks. But in order to visualize the respective difference, the diffusion dynamics of all networks are compared in the subsequent paragraph.

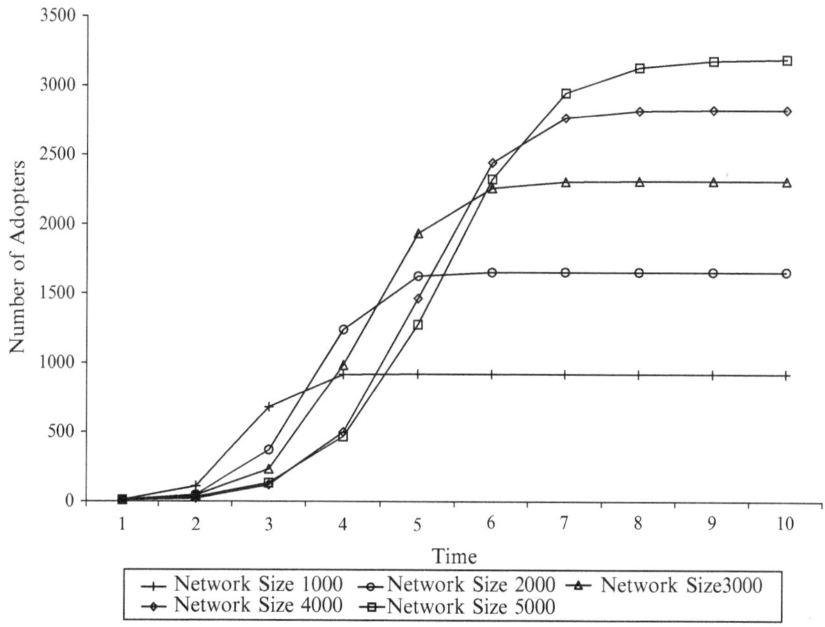

Fig. 12.13 Diffusion dynamics of scale-free networks: Diffusions dynamics in scale-free networks over ten periods
Source: Author

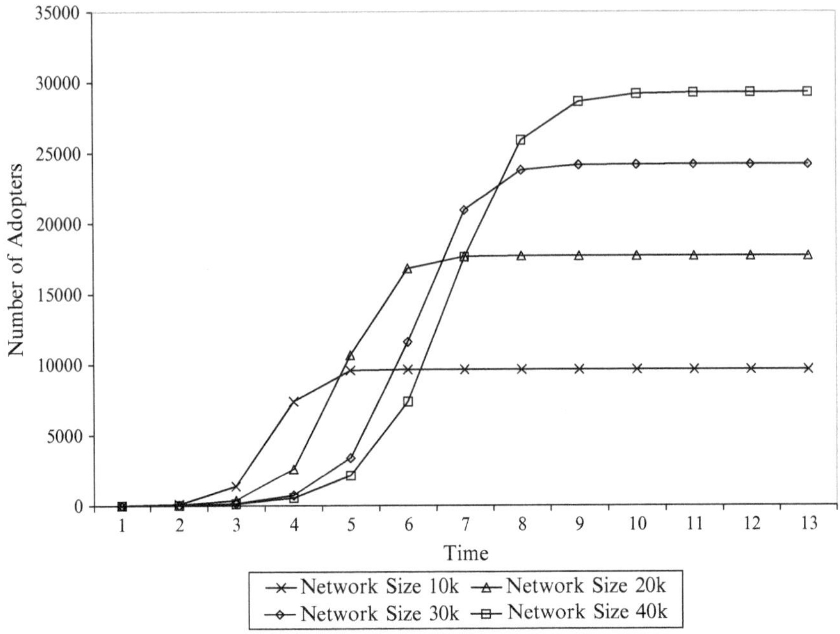

Fig. 12.14 Diffusion dynamics of scale-free networks: Diffusions dynamics in large scale-free networks over 13 periods
Source: Author

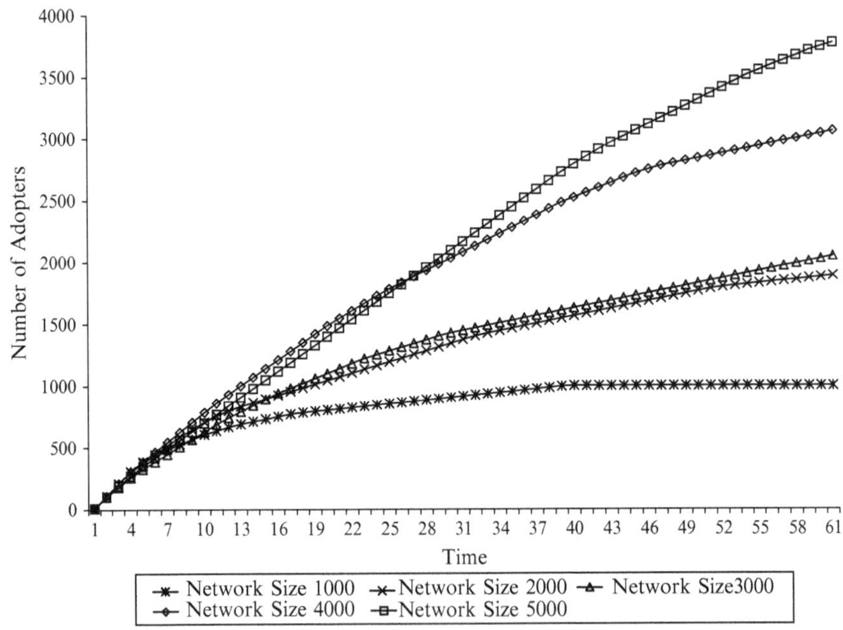

Fig. 12.15 Diffusion dynamics of small-world networks: Diffusions dynamics in small-world networks over 60 periods
Source: Author

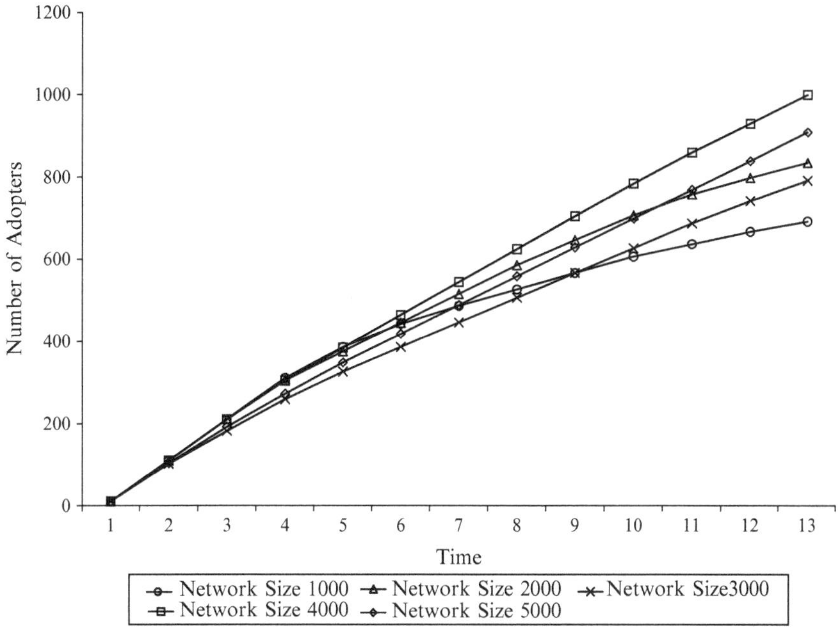

Fig. 12.16 Diffusion dynamics of small-world networks: Diffusions dynamics in small-world networks over 13 periods
Source: Author

5. Summary of Diffusion Dynamics in Software Markets

A comparison of the results from the various models reveals the large differences among the network topologies. In a first step, we compare the all simulations for all network topologies for a very long time horizon. The direct comparison emphasizes the impression based on the first insight. While the diffusion traverses the total population after a very short time at all scales, the diffusion takes longer in scale-free networks and longest in small-world networks. A focus on the first ten periods supports the previous statements. Diffusions in random networks spread faster than in scale-free networks. A comparison of the scale-free topology to small-world networks reveals that diffusion in scale-free networks tend to propagate much faster than in small-world networks, in general. However, it is possible to observe that all diffusions in small-world networks spread faster in period two and three than in scale-free networks (Figs. 12.17 and 12.18).

Moreover, it is possible to aggregate the data in order to derive further conclusions. For this purpose, we calculate the mean of the simulations for all network topology. The resulting graph is depicted in Fig. 12.19. It emphasizes the differences of all network topologies with respect to diffusion dynamics.

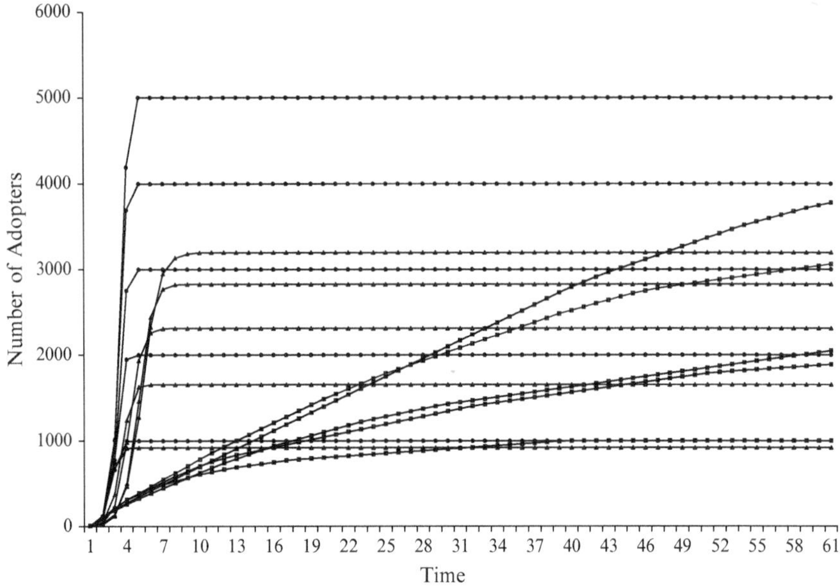

Fig. 12.17 Summary of diffusion dynamics in RN (*circle*), SFN (*triangle*) and SWN (*square*) over 60 periods
Source: Author

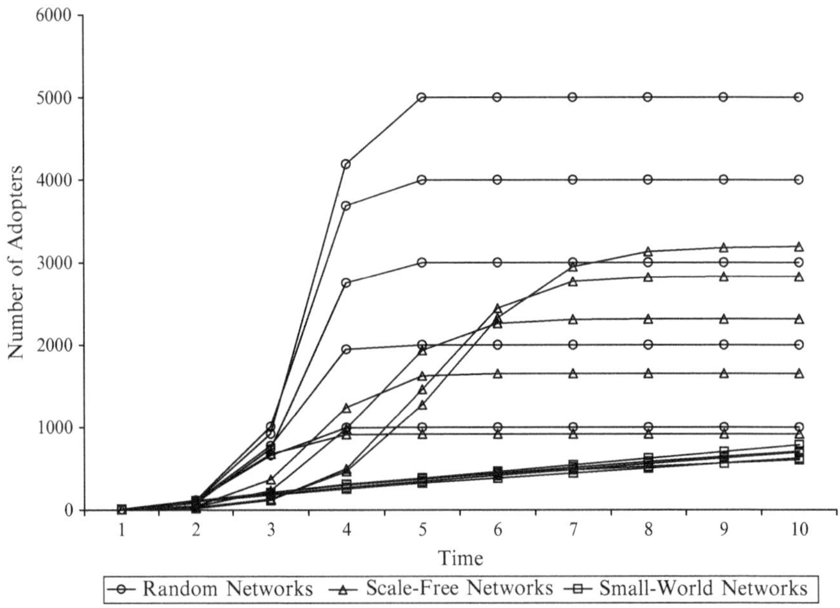

Fig. 12.18 Summary of diffusion dynamics in RN (*circle*), SFN (*triangle*), and SWN (*square*) over ten periods
Source: Author

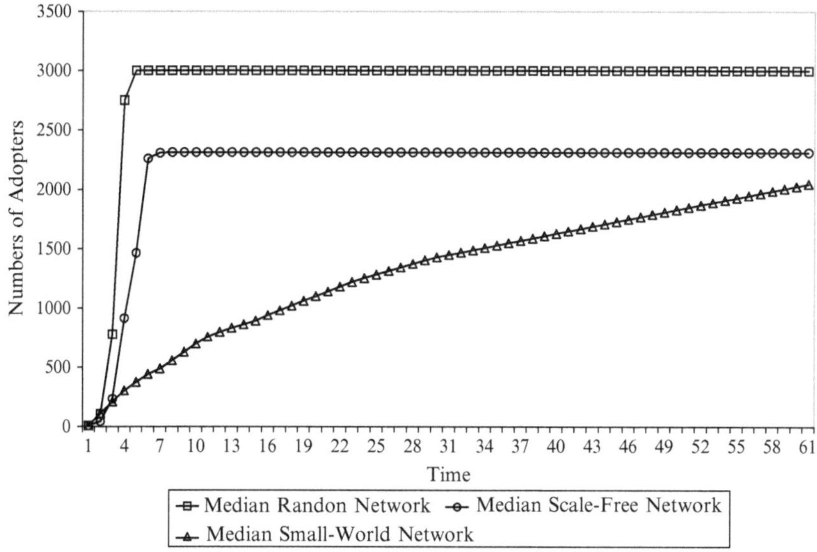

Fig. 12.19 Median of diffusion dynamics over 60 periods
Source: Author

12.1.3 Reconsideration of Diffusion Dynamics In Varying Network Topologies

All in all, the comparison of diffusion dynamics in all relevant network topologies confirms the hypothesis that the variation of the topology has a decisive impact on diffusion dynamics in networks. The simulations and the respective analysis prove by contraposition that the diffusion dynamics of networks are not independent of the underlying network topology. Our results prove that, ceteris paribus, the diffusion dynamics are much faster in scale-free networks than in both other topologies. A comparison of scale-free networks and small-world networks reveals that small-world networks propagated quicker under the given parameters, whereas after the third period diffusions in scale-free networks are much faster than in small-world networks. The respective hypothesis is validated with a prove by contraposition as the findings falsify the hypothesis that diffusion processes are identical in all network topologies. Applied to software markets, this implies that varying network topologies of complex customer networks have an impact on the adoption and diffusion behavior in customer networks of companies operating in software markets. This, in turn, is a decisive insight for understanding network effects in software markets in order to enhance a customer network-centric valuation approach for software markets. Accordingly, it is important to consider network topologies for valuations of software companies. Hence, a respective analysis, e.g., based on the adoption and diffusion simulator, should be integrated into the framework for valuation in software markets. After investigations of the general relevance of network topologies, the following analysis focuses on the scaling properties.

12.2 Scaling Properties of Complex Customer Networks

Previous research revealed the relevance of network topologies for diffusions in complex networks, e.g., complex customer networks. In this section, scaling properties of customer networks are the focus of our investigations. These are conducted as follows. First, the hypothesis is derived and the underlying motivation is explained. Then, the hypothesis is investigated in an analysis based on complementary investigations and simulations. A reconsideration of the most important findings with respect to this hypothesis concludes this section.

12.2.1 Hypothesis on Scaling Properties of Complex Customer Networks

The previous investigations revealed the relevance of network topologies. Therefore, it is possible to consider respective complex network approaches. A popular complex networks approach is to model phenomena and to investigate them with respective simulations, e.g., if full-scale models of large-scale networks cannot be designed due to computational limitations. However, it is important to note, that such an approach requires that networks are invariant to scaling. This invariance to scaling would allow one upscaling, e.g., to increase the size of the customer networks in order to study their growth processes, as well as downscaling, i.e., creating size-reduced models of large scale real-world networks. Initial research efforts suggest that the scale of the network does not have an impact on the properties and dynamics of networks, but the respective hypotheses are not proven (von Westarp 2003). Based on this suggestion, we formulate the hypothesis that relevant properties and dynamics of customer networks in software markets are invariant to scaling [HP2]. It is investigated in the following section.

12.2.2 Analysis of Scaling Properties in Software Markets

Research on the scaling properties of networks reveals a more diverse picture than stated in the hypothesis. The analysis is based on research insights on invariance, mean-field approximation approaches and simulations of networks at various scales. In this discussion, the following aspects are primarily relevant.

Invariance Hypothesis

Researcher investigating social networks in software markets suggest that the scale of the network does not have an impact on its properties and dynamics (von Westarp 2003). But this hypothesis is not proven by experiments or data. It is simply stated that the sensitivity analysis with respect to the network size did not provide an indication that the scale of the network has an impact on the network dynamics (von Westarp 2003). The investigated network consisted of 1,000 agents which was a considerable size at that time with respect to the available computing power. Today, the advanced computational power and network design allow one to simulate much larger networks if necessary.

Mean-Field Approximations

Another school of thought investigates the same problem from a slightly different angle as the underlying adoption rule is based on a mean-field approximation. Following a mean-field approximation, the dynamics of networks will only be independent of system size if the neighborhood increases on the same order of magnitude as the system size (Alfarano and Milakovic 2008). A comparison of all network topologies reveals that only random graph networks are scalable under the condition that the average connectivity is proportional to the size of the network. Equivalently, the respective link probability has to be kept constant. More general, any network topology fulfilling this relation would be invariant to scaling. However, this relationship doesnot hold for small-world and scale-free networks if the underlying adoption rule is based on a mean-field approximation. However, it is important to note that these findings do not necessarily have an impact on the validity of our simulator as the outlined adoption rule is different.[14]

Simulation of Diffusions at Various Scales

In addition to the previous investigations, simulations of diffusion process at various scales are conducted and compared to each other. For this purpose, the complex networks adoption and diffusion simulator is applied in order to examine diffusions all else equal at various scales in all relevant network topologies. The underlying rationale of these investigations is simple. If the networks are not invariant to scaling, iterations of the same adoption rule with identical parameters applied at various scales would calculate different results. In the following, the respective findings are structured according to network topologies before they are compared in the final paragraph of this subsection.

1. General Diffusion Process and Parameters

Research on the financial herding model outlines the importance of the average linking probability in random networks (Alfarano and Milakovic 2008). Although this

[14] Please confer Chap. 11.

model does not match entirely our complex networks adoption and diffusion simulator due to different adoption rules, it is possible to transfer the underlying idea to our investigations of customer networks in software markets. Accordingly, simulations of scaling properties of random network with the developed complex networks adoption and diffusion simulator illustrate that some network topologies are invariant to scaling. A comparison of adoption and diffusion processes at various scales of random networks reveals that similar percentages of adopters can be observed for changing sizes of the customer networks. However, it is important to note that this scaling invariance depends on an average linking probability that is kept constant, by modifications of the average connectivity accordingly to the increasing scale of the network. In order to prove these derivations with own simulations, the following investigations are conducted ceteris paribus in various network topologies and on various scales. All simulations of the upcoming investigations share the following assumptions:

1. *Network topology.* We assume the three most relevant network topologies in the upcoming subsections. This is the only parameter that is changed in the respective analysis.
2. *Network size.* The size of the population is changed in order to compare the results of the diffusion process all else equal at different scales. We conduct investigations at various scales of observation. The standard size of the investigated networks range between 10,000 and 40,000 with intervals of 10,000.[15]
3. *Adoption rule.* We assume in all models the previously presented decentralized standardization rule based on the underlying assumptions.[16]

2. Diffusion Results in a Random Network

In the first step, a small random network N_S is created that has $5,001$ nodes and an average connectivity of 5. As the average linking probability p is defined as

$$p = \frac{D}{N-1}, \quad (12.1)$$

with network size N and average connectivity D, the average linking probability of the small network p_S is equal to

$$p_S = \frac{5}{5{,}001 - 1} = 0.001. \quad (12.2)$$

[15] Please note that the simulator provides also the possibility to implement stochastic distributions of utilities in order to simulate heterogeneous agents, but such investigations are is beyond the scope of this book.

[16] Please confer Sects. 7.2.2.2 and 11.4.2 for a description of its algorithmic implementation.

Accordingly, the customer adoption rule based on the decentralized standardization model is parameterized with the following input. Set-up costs are assumed to be 10, the average direct utility is set to 5, the indirect utility is set to 1 and the number of initial adopters is set to 50. There are no variations of the direct utility assumed, no direct costs, no indirect costs and no variation of the indirect benefit. After six iterations nearly all customers, to be exact $4,958$ of the 5,000 representing 99% of the total population, have adopted the software. In the second step, a large random network N_L is created that has $50,001$ nodes. Hence, it scales at a factor 10 in comparison to the previous smaller network. As the average linking probability of the large random network p_L is kept constant rearrangements of

$$p = \frac{D}{N-1}, \tag{12.3}$$

for D reveal a required connectivity for the large random network of

$$D = p \cdot (N-1) = 0.001 \cdot 50,001 - 1 = 50. \tag{12.4}$$

The adoption rule is still based on the decentralized standardization model and parameterized with the following input parameters. Set-up costs C_S are assumed to be 10, the average direct utility $U_{d}ir$ is set to 5, the indirect utility is set to 1 and the number of initial adopters is set to 500. There are no variations the direct utility assumed, no direct costs, no indirect costs and no variation of the indirect benefit. After a few iterations, similar to the adoption and diffusion process of the small network nearly all customers, to be exact 99% of the total populations, have adopted the software. This suggests that random networks are scaling invariant as other iterations with similar relationships delivered similar results.

In addition, we conduct further, more systematic simulations. This time the size of the random networks range from 10,000 to 40,000 in steps of 10,000. Based on the outlined assumptions, the complex networks adoption and diffusion simulator is used. The results of our simulations on varying scale random networks are depicted in Fig. 12.20. It summarizes the diffusion dynamics for the simulated networks for 11 periods. The plot reveals again that diffusion processes in random networks are very fast. After six iterations even a network of size 40,000 is percolated. Please note the difference of adopters between period three and four.

3. Diffusion Results in a Scale-Free Network

The same diffusion process is modeled on scale-free networks. Our findings are depicted in Fig. 12.21. If the same diffusion process takes places on a scale-free network, the following findings are derived based on the previously stated assumptions with respect to the diffusion process. The results reveal that diffusions in large scale-free networks occur still very fast. A maximum of nine iterations is required in

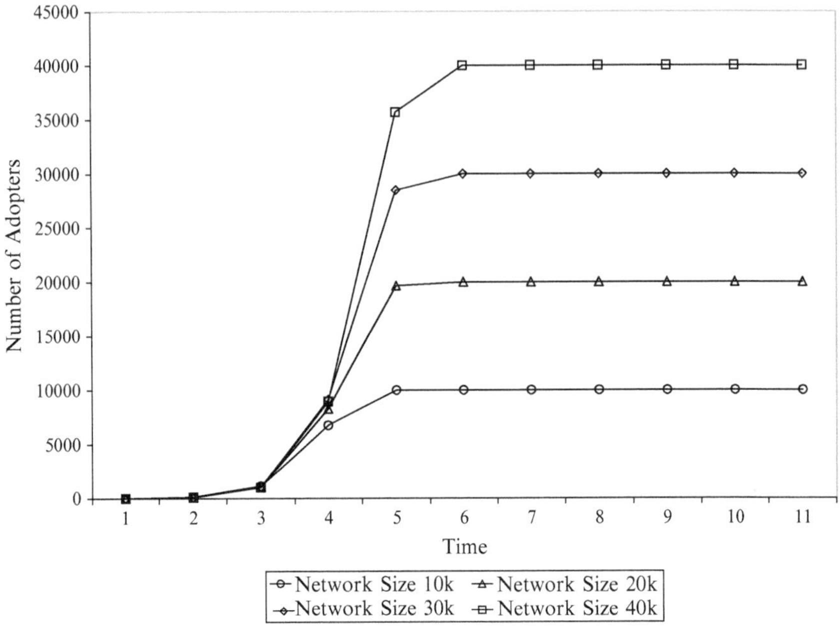

Fig. 12.20 Scaling in random networks for 11 periods
Source: Author

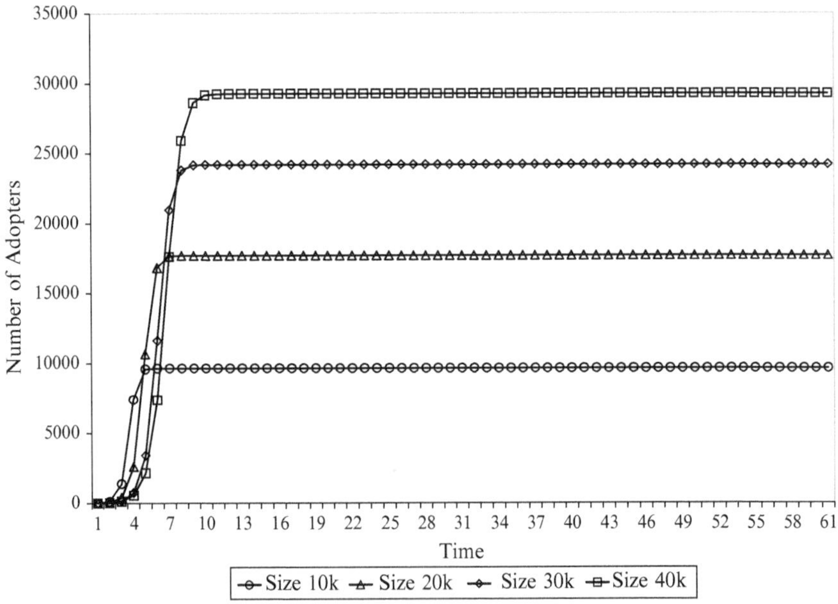

Fig. 12.21 Scaling in scale-free networks for 60 periods
Source: Author

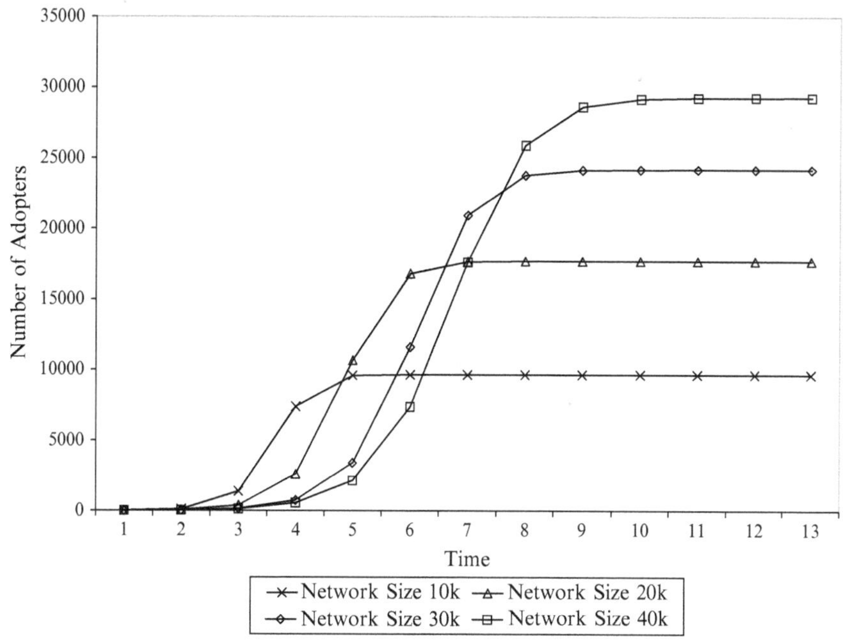

Fig. 12.22 Scaling in scale-free networks for 13 periods
Source: Author

order to traverse the total network. A more detailed analysis of the first 13 iterations is summarized in Fig. 12.22

4. Diffusion Results in a Small-World Network

The simulator was used to simulate the same diffusion process on small-world networks. Results are summarized in Fig. 12.23.

5. Summary of Diffusion Dynamics at various Network Scales

A comparison of the results from the various models reveals that the diffusion dynamics of networks depend on the underlying network topology. All else equal, the diffusion reaches much faster people in random networks than in scale-free networks. In addition, the our results indicate that diffusion in scale-free networks are much faster than in small-world networks. All findings illustrate the crucial impact of structural heterogeneity on the distributional outcome of the model. Random networks are the only prototypical network structure that is invariant to scaling. An underlying assumption is that we restrict the description of network characteristics to homogeneous links and nodes. Such investigations reveal that structural heterogeneity matters for macroscopic properties of the network (Figs. 12.24 and 12.25).

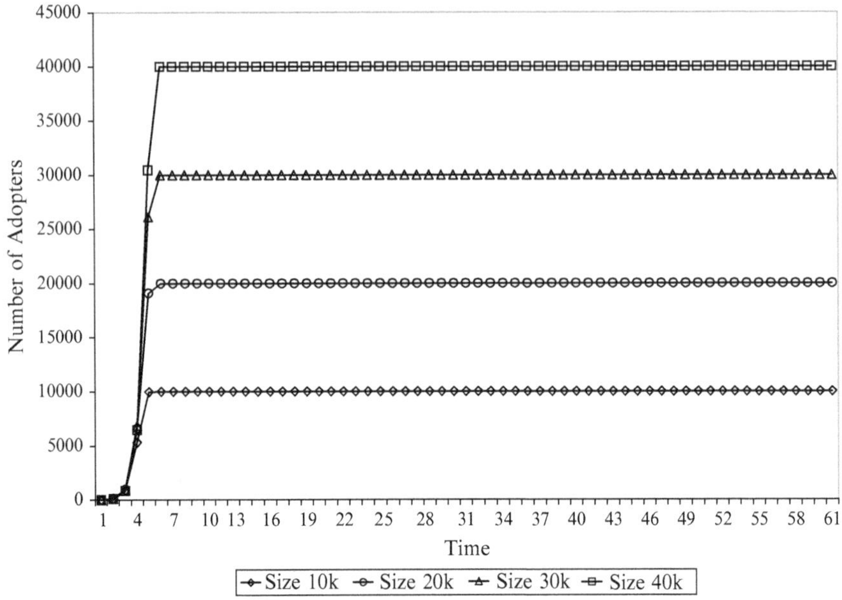

Fig. 12.23 Scaling in small-world networks for 60 periods
Source: Author

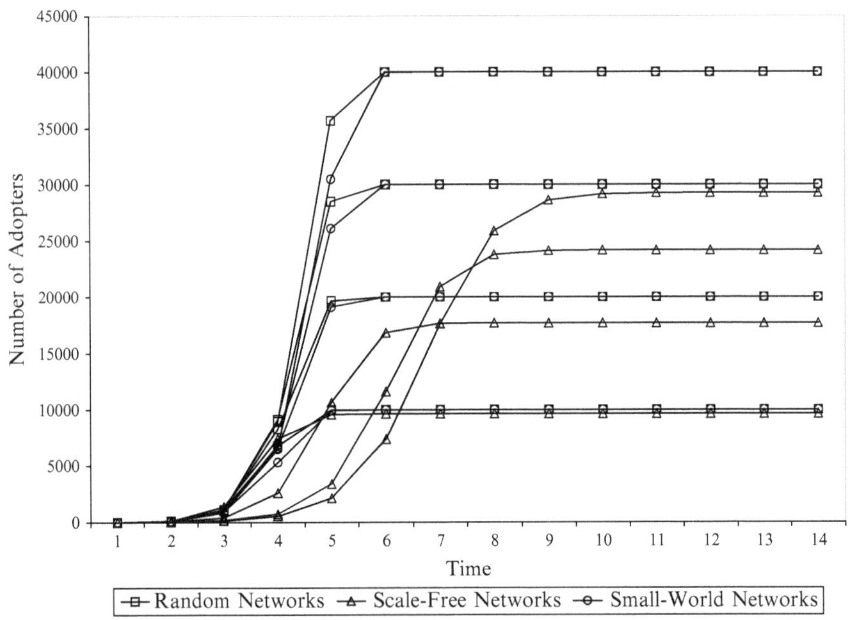

Fig. 12.24 Comparison of scaling properties for 14 periods
Source: Author

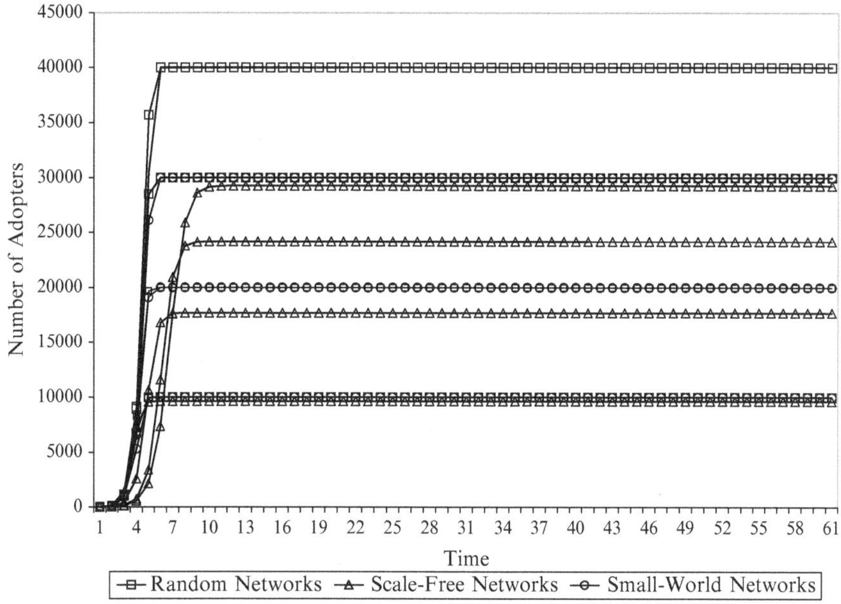

Fig. 12.25 Comparison of scaling properties for 60 periods
Source: Author

12.2.3 *Reconsideration of Scaling Properties of Complex Customer Networks*

Previous research efforts suggested that the scale of a network does not have an impact on its properties and dynamics. But the conducted theoretical and numerical investigations concerning scaling properties of complex customer networks revealed a more differentiated picture. In essence, the scaling properties of network depend on the underlying network topology. Depending on their topology, networks exhibit different characteristics at various scales of observation. Additional investigations with the adoption and diffusion simulator support this differentiated picture. For this reason, the general form of the hypothesis is rejected. Instead, we conclude that random networks are invariant to scaling, whereas scales are relevant in scale-free networks and small-world networks. Thereby, the conducted research fills a research gap with a counter-intuitive result.

12.3 Network Topologies of Complex Customer Networks in Software Markets

Since our previous research revealed that network topologies are relevant for customer networks in software markets, it is a primary research interest to specify their exact class. Hence, the following three steps are conducted. First, the hypothesis is formulated and its motivation is depicted. In a second step, the hypothesis is investigated in an analysis based on complementary investigations and simulations. Finally, the most important findings are summarized in the reconsideration of the hypothesis.

12.3.1 Hypothesis on the Topology of Customer Networks in Software Markets

As network topologies are relevant in customer networks of software markets, their exact nature is investigated. Traditional research frequently assumes implicitly fully mixed networks, e.g., due to analytical descriptions, which is equivalent to the implicit assumption of a random network topology. Complex networks research, in turn, reveals that most real-world social networks tend to be in general small-world networks (Watts 1999; Albert 2001; Barabasi et al. 2000). In order to resolve this discrepancy, the nature of customer networks in software markets is investigated. For this purpose, we frame the hypothesis based on complex networks insights that complex customer networks in software markets also tend to be small-world networks [HP3]. This hypothesis is of primary interest as a confirmation of the hypothesis would allow one to transfer the respective network theoretical insights concerning small-world networks to complex customer networks in software markets.

12.3.2 Analysis of the Topology of Customer Networks in Software Markets

The investigations of the hypothesis are conducted by analyzing the findings of complex networks research on relevant Small-World Network Analyses in social networks, by observations on opinion leader in social networks, and by a supportive case study of the German Xing AG.

Small-World Network Analyses of Social Networks

A review of the relevant complex networks literature supports the hypothesis that customer networks are primarily small-world networks, although most conventional models assume implicitly that customer networks are rather random networks. First indications that social networks are primarily small-world networks date back to experiments on the 6° of separation. Empirical research on the topology of social networks revealed that the diameter of the corresponding graph of social connections is not much larger than six (Milgram 1967; Travers and Milgram 1969;

Garfield 1979; Guare 1990). Additional research supports the general insight that for some complex networks only a short path is needed to connect even the most distant members (Sattenspiel and Simon 1988). Additional supportive research is provided by Watts and Strogatz who investigated a whole variety of social networks (Watts and Strogatz 1998; Watts 1999). Moreover, there is another school of thought with a focus on business communities, such as intercorporate networks. Accordingly, the relationships between management boards of all industries are small-world networks (Mariolis 1975; Galaskiewicz and Marsden 1978; Mizruchi 1982; Davis and Greve 1997; Kogut and Walker 2003). In addition, there are reviews on relevant network metrics of various social networks, e.g., citation network and actor networks (Albert 2001; Newman 2003b). Figure 12.26 contains information on the following parameters of social networks:

	network	type	n	m	z	l	a	$C^{(1)}$	$C^{(2)}$	r
social	film actors	undirected	449 913	25 516 482	113.43	3.48	2.3	0.20	0.78	0.208
	company directors	undirected	7 673	55392	14.44	4.60	–	0.59	0.88	0.276
	math coauthorship	undirected	253 339	496 489	3.92	7.57	–	0.15	0.34	0.120
	physics coauthorship	undirected	52 909	245 300	9.27	6.19	–	0.45	0.56	0.363
	biology coauthorship	undirected	1520 251	11 803 064	15.53	4.92	–	0.088	0.60	0.127
	telephone call graph	undirected	47 000 000	800 000 000	3.16		2.1			
	email massages	directed	59 912	86 300	1.44	4.95	1.5/2.0		0.16	
	email address books	directed	16 881	57 029	3.38	5.22	–	0.17	0.13	0.092
	student relationships	undirected	573	477	1.66	16.01	–	0.005	0.001	–0.029
	sexual contacts	undirected	2810				3.2			
information	WWW nd.edu	directed	269 504	1 497 135	5.55	11.27	2.1/2.4	0.11	0.29	–0.067
	WWW Altavista	directed	203 549 046	2 130 000 000	10.46	16.18	2.1/2.7			
	Citation network	directed	783 339	6 716 198	8.57		3.0/–			
	Roget's Thesaurus	directed	1 022	5 103	4.99	4.87	–	0.13	0.15	0.157
	word co–occurrence	undirected	460 902	17 000 000	70.13		2.7		0.44	
technological	Internet	undirected	10 697	31 992	5.98	3.31	2.5	0.035	0.39	–0.189
	power grid	undirected	4 941	6 594	2.67	18.99	–	0.10	0.080	–0.003
	train routes	undirected	587	19 603	66.79	2.16	–		0.69	–0.033
	software packages	directed	1 439	1 723	1.20	2.42	1.6/1.4	0.070	0.082	–0.016
	software classes	directed	1 377	2 213	1.61	1.51	–	0.033	0.012	–0.119
	electronic circuits	undirected	24 097	53 248	4.34	11.05	3.0	0.010	0.030	–0.154
	peer–to–peer network	undirected	880	1 296	1.47	4.28	2.1	0.012	0.011	–0.366
biological	metabolic network	undirected	765	3 686	9.64	2.56	2.2	0.090	0.67	–0.240
	protein interactions	undirected	2 115	2 240	2.12	6.80	2.4	0.072	0.071	–0.156
	marine food web	directed	135	598	4.43	2.05	–	0.16	0.23	–0.263
	freshwater food web	directed	92	997	10.84	1.90	–	0.20	0.087	–0.326
	neurel network	directed	307	2 359	7.68	3.97	–	0.18	0.28	–0.226

Fig. 12.26 Overview complex networks studies
Source: (Newman 2003b)

n Number of nodes
m Number of edges
z Mean degree
l Diameter
α Exponent of degree distribution if it follows a powerlaw
C^1 and C^2 Clustering Coefficients
r Degree correlation coefficient

A Small-World Network Analysis of these social networks considers primarily the clustering coefficient and the characteristic average path length or the diameter.[17] Many social networks have a relatively high clustering coefficient in combination with a low diameter. The only significant exceptions are student relationships, which are characterized by a very high diameter in combination with a very low clustering coefficient. This can be explained by the research design with a very small sample size of 573 network participants in combination with a very low number of links 477. All in all, the studies reveal that social networks tend to be small-world networks.

More recent research investigating the structure of social networks reveals that the average path length of messenging networks, such as the Microsoft Messenger, is 6.6. This is close to the 6° of separation (Leskovec and Horvitz 2008). The study is based on anonymized data that contains 30 billion conversations among 240 million people capturing a month of high-level communication activities within the Microsoft Messenger instant-messaging system. A network theoretical analysis illustrates that the graph of the users is well-connected, robust to node removal and that the average path length among Messenger users is 6.6 (Leskovec and Horvitz 2008). The findings are in line with similar studies on email networks (Dodds et al. 2003).

In summary, it is possible to conclude that social networks tend to be small-world networks. As customer networks in software markets are typical social networks, it can be deduced that customer networks in software markets tend also to be small-world networks.

Opinion Leader in Complex Customer Networks

The literature on opinion leaders in social groups such as customer networks illustrates that some agents can be far more persuasive than others (Valente and Davis 1999). This heterogeneity of the customers is important. With respect to software markets the presented technology acceptence model reveals that customer networks in software markets have a small-world.[18]

Case Study Xing AG

A case study on the Xing AG, a German business network, reveals that social contacts in customer networks are organized as small-world networks. Unfortunately, there are only few information available on the customer network of Xing. But the company revealed the historical development of the customer network in various presentations. Unfortunately, there is no public information available on the degree distribution of the customer network. The annual customer survey revealed that the average user had 26 contacts in 2005. This number increased to fifty confirmed

[17] Please confer Sect. 10.3.2.

[18] Please confer Sect. 7.2.1.

contacts per user in 2006 (Xing 2007a,b). From a complex networks perspective, the average connectivity of the network increased from 26 to 50 during this time period. It increased further to 103 in 2007 (Xing 2007b). A key information is that more than every sixth contact has more than 150 contacts. Additional research on the customer network of Xing reveals, that the average number of edges between any two vertices is very small. This can be derived by comparing the number of connections to the number of secondary and tertiary connections. The clustering coefficient, however, is large. There are some short cuts in the Xing network, such as member groups, which decrease the average path lengths of the customer network. The combination of a relatively high clustering coefficient with a low characteristic path length supports the hypothesis that the customer network is a small-world network. Hence, the case study on the Xing AG supports the hypothesis, that customer networks tend to be small-world networks.

12.3.3 Reconsideration of Topology of Customer Networks in Software Markets

Since network topologies are relevant to customer networks of software markets, the next logical step was to specify their exact nature. The Small-World Network Analysis confirms that complex customer networks in software markets are primarily small-world networks. This finding is support by supportive research on other social networks and the presented case study. At the same time, our findings reveal that it is reasonable to transfer insights on small-world networks to customer networks topology in the subsequent investigations.

12.4 Contributions to Customer Network-Centric Valuation

In this section, we investigate possible contributions from complex networks research to customer network-centric valuation. For this purpose, the following steps are conducted. In a first step, the hypothesis is formulated and motivated. In a second step, the hypothesis is investigated in an analysis based on complementary investigations and simulations. Finally, the most important findings are summarized in the reconsideration of the hypothesis.

12.4.1 Hypothesis on Contributions to Customer Network-Centric Valuation

The primary goal of all conducted investigations is to enhance the quality of customer network-centric valuation in software markets. Hence, we derive the hypothesis that the complex networks tools can be applied to customer networks in software markets in order to obtain supportive information on valuation parameters if the required information on the underlying customer network is accessible and reliable [HP4]. This hypothesis is a fundamental link connecting the previous research on corporate valuation in software markets, network effects and the outlined complex networks research.

12.4.2 Analysis of Contributions to Customer Network-Centric Valuation

Previous investigations of the relevant valuation approaches emphasize the vital importance of two parameters. Those of the price of the underlying and of the volatility are vital parameters for valuation in software markets. Hence, it is important to assess the possible insights that can be derived from a complex networks perspective with respect to both parameters.

Complex Networks Information as Contributions to Customer Network-Centric Valuation

Research reveals that complex networks and complex networks contribute to customer-network centric valuations based on the following reasoning. In a first step, it is important to note that the customer base is an important value driver of software companies. It has been noticed that software purchasing decisions are significantly influenced by network effects, e.g., the number of communication partners using the very same software or the total number of licenses sold. Hence, adoptions and diffusions of products are key driving forces of successful business models in software markets. They determine the development of customer networks which are frequently the only significant assets of software companies. In this light, the knowledge of dynamical and statistical properties about product diffusions in software markets is of fundamental importance in order to assess business models, valuations and subsequent investment decisions. They provide additional information for the valuation as depicted in the following sections.

Derivation of the Price of the Underlying

From a network theoretical perspective, the assessment of the customer network is at the heart of the dilemma to project the expected cash flows. A vital parameter of all relevation valuation approaches is to determine the *expected cash flows*. Future cash flows are frequently vague, as they depend on sales projections. From a complex networks perspective, such projections, in turn, are traditionally based on implicit assumptions about the underlying customer network. In practice, sales projections are frequently adjusted by linear approximations of the historical corporate performance. But as customer networks are subject to complex network dynamics, even small impulses can have a vital impact on the development of the customer network and of the company performance. Hence, an explicit treatment of the network effects is required. If customer networks are interpreted as the underlying of a software company the complex networks investigations of the customer networks may contain additional information in order to derive better approximations of the input parameters and to increase, thereby, the quality of the valuation.

In this research context, network theoretical investigations of critical masses in customer networks are highly relevant.[19] In general, the critical mass is defined as the minimum scale of operation that is required for sustainable business models. In terms of network theory, the critical mass of a network can be analyzed by investigating phase transitions of its giant component.[20] Accordingly, critical masses of networks can be approximated as a plot of the giant component reveals a nonlinear phase transition if the critical transition boundary is crossed. Above the critical transition boundary, the diffusion of the is locked-in. If, in contrast, the corporation is not capable to pass the transition boundary, rebalancing feedback dynamics are likely to drive the product out of the market.

The application of the complex networks adoption and diffusion simulator is one way to quantify the critical mass of a network and to derive, thereby, a reliable proxy for the development of the underlying. In the first step, the set of all potential customers is interpreted as a customer network and the parameters of the model are derived from empirical statistics. Accordingly, the infection rate represents the probability that a customer purchases a product because of an infection from a network neighbor. The infection rate itself is effected by the marketing strategy, the need for the product, and other market parameter. The recovery rate, in turn, is equal to the sum of the effects that might lead to an end of the consumption, e.g., force of substitution, competition and others. Once, the parameters of the model are determined, numerical simulations based on the adoption and diffusion simulator can be applied in order to determine the respective distribution of customers in the investigated customer network. In combination with the assumptions concerning other financial parameters, expectations can be formulated concerning the overall revenues of the customer network. Once the revenues of the assessed investment are approximated, the network effects on other vital valuation input parameters have to be assessed in order to finally derive the underlying cash flows of the corporation. If the model parameters are varied within the most realistic regimes, it is possible to calculate the distribution of the investment cash flows for different scenarios around the mean and the corresponding volatility.[21]

Once the volatility, the price of the underlying, the strike price, and the dividend payments are known, the complementary option value parameters time to

[19] Please note that the following investigations are based on the presented complex networks research, i.e., epidemiology, percolation, and phase transitions. Please confer Sect. 10.5.

[20] Please note that the critical mass is not necessarily identical to the financial break-even of a company. Therefore, it is determined rather by topological properties and dynamics of the customer network rather than by the current economic performance of a company.

[21] Please confer the next section for further information on the implentation of such an approach.

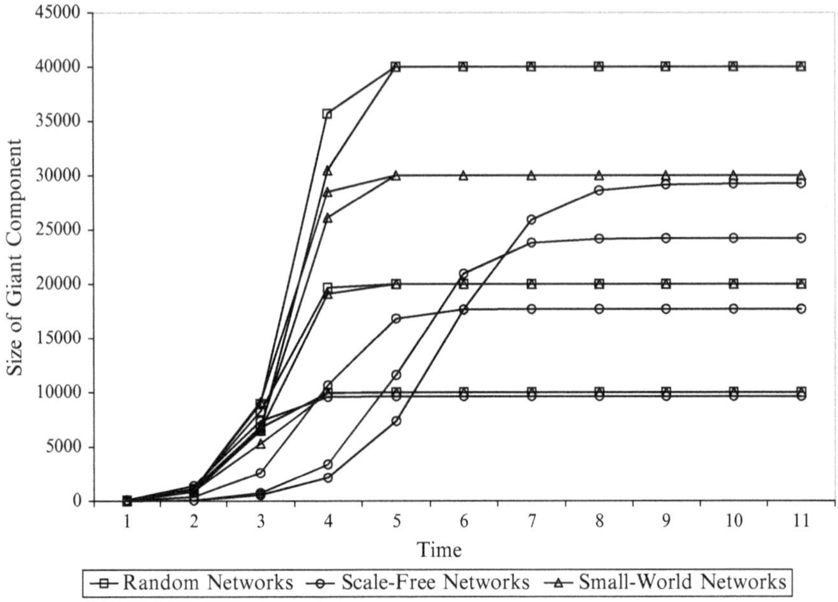

Fig. 12.27 Critical masses of complex networks quantified by their giant component
Source: Author

maturity and interest rate have to be determined for the idiosyncratic valuation context, in order to calculate a nominal value range by applying the option pricing methodology. The vital added value of the internalization of network effects is the incorporation of additional information on the customer network. Since customer network dynamics are governed by discontinuous phase transition similar to other social networks, their cash flows are not normally distributed as in a conventional Monte-Carlo simulation. In contrast, the cash flow distributions reveal a discrete jump for the emerging giant component above the phase transition boundary. For this purpose, the adoption and diffusion simulator can be applied in order to investigate the giant component of networks. It is important to note the difference between all customers and the members of the giant component. While the number of adopter provides information on the overall network, the giant component is an important subset. It is the largest cluster of connected customers. Therefore, the development of the giant component allows one to identify and to quantify critical masses of networks.

The plot of a giant component allows one to identify a range of customers that are required in order to incite a nonlinear increase of customers.[22] The following example illustrates our reasoning. Figure 12.27 is a plot of various giant components

[22] Please note that the giant component decreases also exponentially if a central hub is removed. Such investigations are the research focus of network resilience analyses which is beyond the scope of this book.

that result from our experiments with random networks, scale-free networks and small-world networks. In summary, the presented approach provides some decisive advantages in comparison to conventional approaches. Consequently, the additional information improves the approximation of the cash flow distribution and ameliorates the results of the valuation. All in all, the outlined findings support the hypothesis that the innovative approach provides additional information on complex customer networks which are relevant to valuation in software markets.

Derivation of the Volatility

The determination of the volatility is a crucial but also very difficult step in real option valuation. At the same time, research reveals that there are a variety of traditional methods to derive the volatility, such as stochastic process models or monte-carlo simulations. The crucial issue remains whether these tools are good approximations particularly since they assume implicitly random networks and do not account for topological information. Based on the previous insights on networks effects in customer networks, we present an approach to derive the volatility by accounting also for the topological interdependencies. The network theoretical investigations are conducted in the following steps.

1. Simulations of customer network developments
2. Determination of a frequency table
3. Derivation of volatility

This framework is applied, ceteris paribus, in the following, in order to simulate the development of customer networks in random networks and in small-world networks. For this purpose, the following research is pursued.

1. Simulations of Customer Network Developments RN

The adoption and diffusion simulators is used in order to simulate various developments of the customer network.[23] Please note that this random network was generated with two stochastic components in order to generate a range of possible outcomes. We depict the most relevant developments of the diffusion process in the following Fig. 12.28. It summarizes the diffusion dynamics of the simulated small-world network for nine periods. In a next step, the range of possible values for the customer network development for time $t = 4$ is summarized in Fig. 12.29.

[23] Parameter for Random Network: Size: 1,000; Connectivity: 10; Beta: 0,5; Initial Purchasing Costs: 100; Direct Utility: 50, Variance of Direct Utility: 50; Indirect Utility: 10; Variance of Indirect Utility: 10 and Initial Adopters: 10

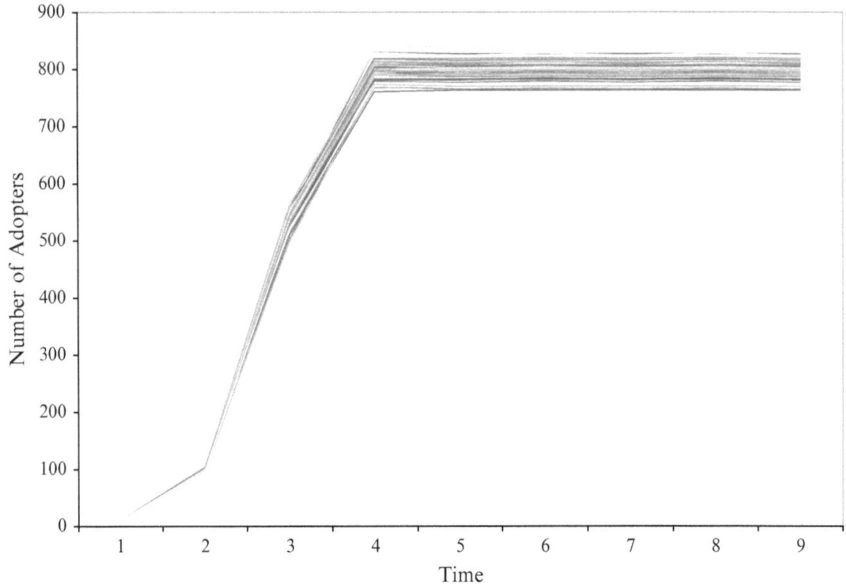

Fig. 12.28 Possible customer network developments for nine periods RN
Source: Author

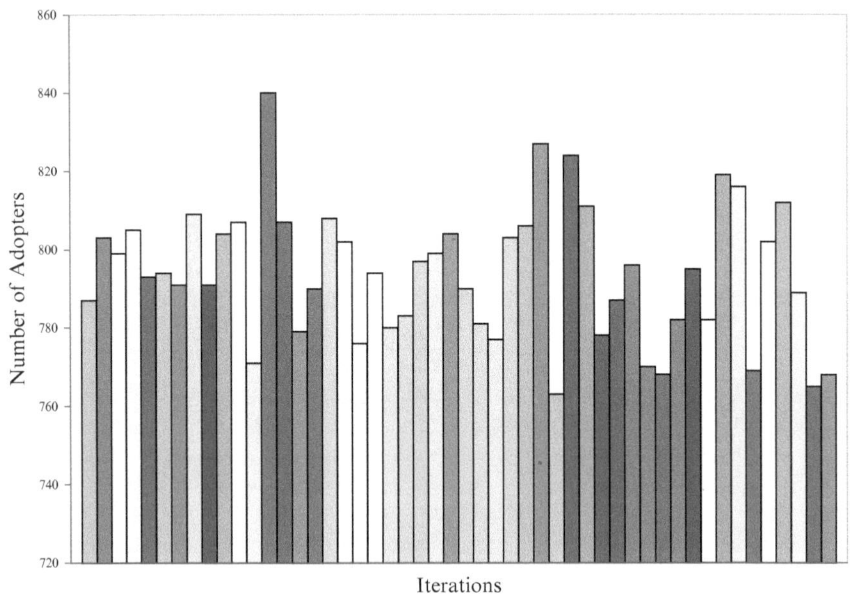

Fig. 12.29 Range of customer network developments in $t = 4$
Source: Author

2. Determination of a Frequency Table RN

Based on the values of the simulation a frequency table is derived for $t = 4$. A frequency table is constructed by dividing the values for possible customer network developments into intervals and by counting the number of scores in each interval. For this purpose the values are clustered based on limit values that determine the frequency intervals. Accordingly, we calculate the average of all iterations. The conducted simulations have an average of 793.86 in period four. Then, the limit values of the frequency classes is determined. We set the interval boundaries to 3.5%. In a next step, the results of the simulation are clustered with respect to the derived limit values into frequency classes, which, in turn, are stated as a percentage of the total observations. Then, the limit values can be depicted in a frequency distribution diagram with the respective probabilities. Please confer Fig. 12.29 for the frequency table for the conducted simulations.

3. Derivation of Volatility RN

The mean μ and the volatility σ of the random network can be derived. For the given values, the respective volatility is 20.1%. In a next step we simulate the same diffusion process, ceteris paribus, on a small-world network. While traditional research approaches assume frequently random network structures, our investigations indicate that customer networks tend to be small-world networks. Hence, we compare diffusion processes in both networks. The purpose of this comparison is to illustrate the difference between results in random network and small-world network.

4. Simulations of Customer Network Developments SWN

The adoption and diffusion simulator is used in order to simulate a small-world customer network with identical network and diffusion parameters.[24] Please note that the randon network was also generated with two stochastic components. The most relevant developments of the diffusion process are depicted in Fig. 12.28. It summarizes the diffusion dynamics of the simulated small-world network for nine periods (Figs. 12.30 and 12.31).

The range of possible outcomes for the customer network development at time $t = 4$ is presented in Fig. 12.32. This data can be now transformed into a frequency table. Hence, we apply the required steps in the following paragraph.

---

[24] Parameter for Small-World Network: Size: 1,000; Connectivity: 10; Beta: 0,5; Initial Purchasing Costs: 100; Direct Utility: 50, Variance of Direct Utility: 50; Indirect Utility: 10; Variance of Indirect Utility: 10 and Initial Adopters: 10

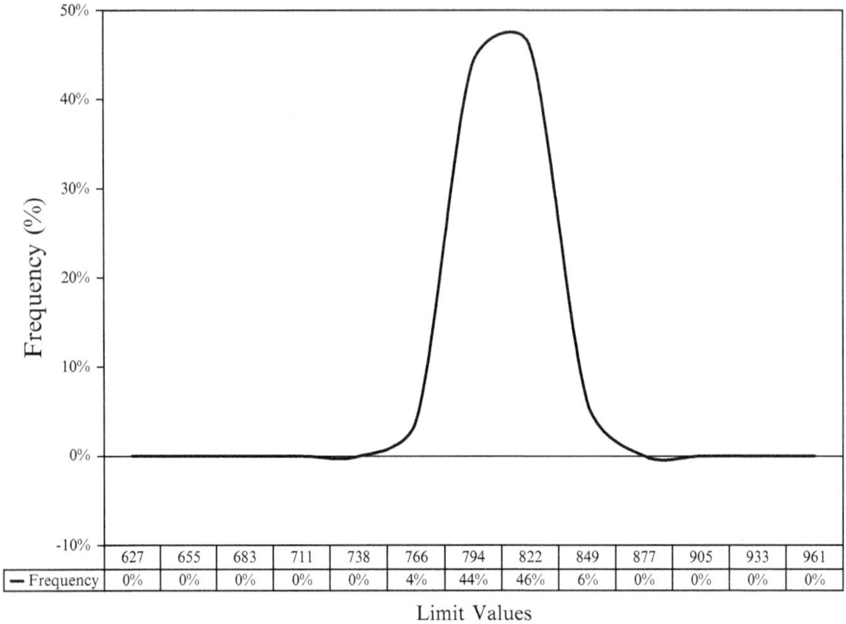

Fig. 12.30 Frequency table for possible customer network developments in $t = 4$
Source: Author

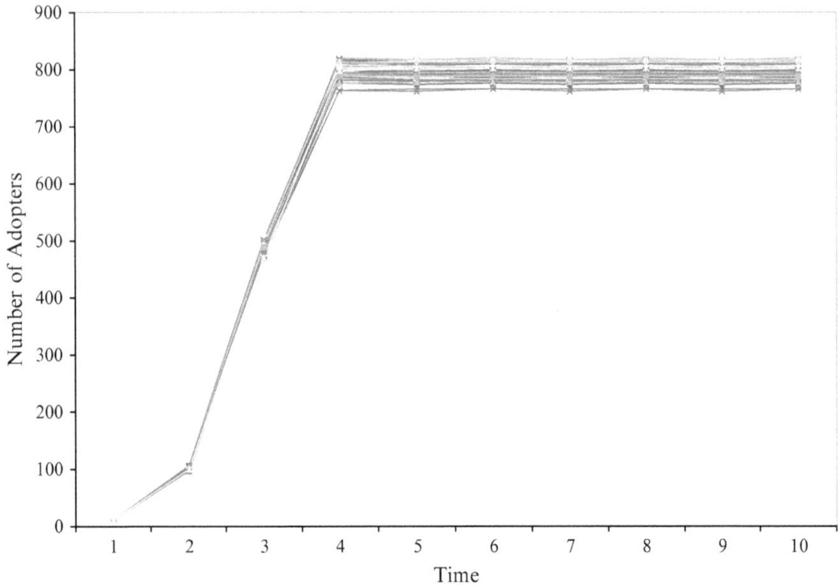

Fig. 12.31 Possible customer network developments for nine periods SWN
Source: Author

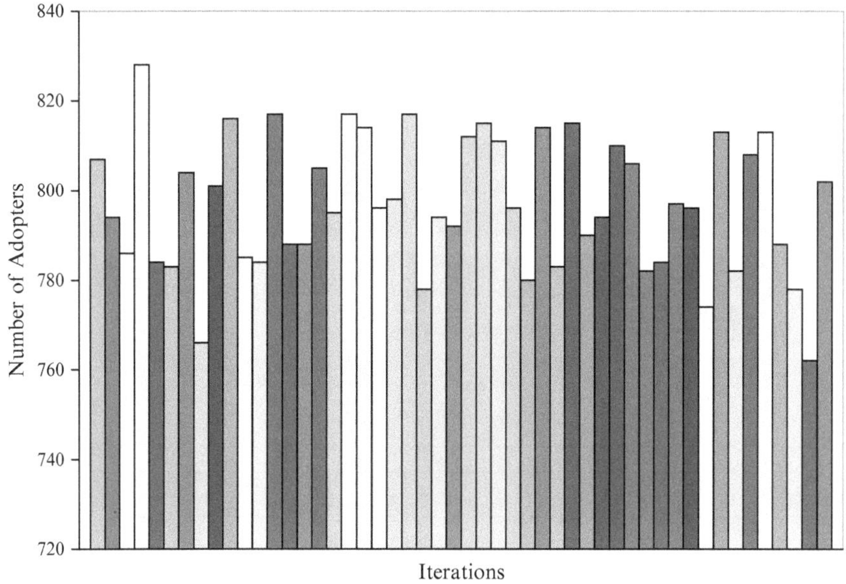

Fig. 12.32 Range of customer network developments in $t = 4$ (SWN)
Source: Author

5. Determination of a Frequency Table SWN

In a next step, is investigated. For this purpose, the required limit values is clustered based on limit values that determine the frequency intervals. Accordingly, we calculate the average of all iterations. The conducted simulations have an average of 793.86. Then, the limit values of the frequency classes is determined. We set the interval boundaries to 3.5%. In a next step, the results of the simulation are clustered with respect to the derived limit values into frequency classes, which, in turn, are stated as a percentage of the total observations. Then, the limit values can be depicted in a frequency distribution diagram with the respective probabilities. Please confer Fig. 12.29 for the frequency table for the conducted simulations (Fig. 12.33).

6. Derivation of Volatility SWN

Then, the mean μ and the volatility σ can be derived.

7. Comparison of results

Our previous research revealed that customer networks tend to be small-world networks. Hence, we compare the diffusion processes in both networks. The comparison reveals that the network topology of the customer network has an impact on

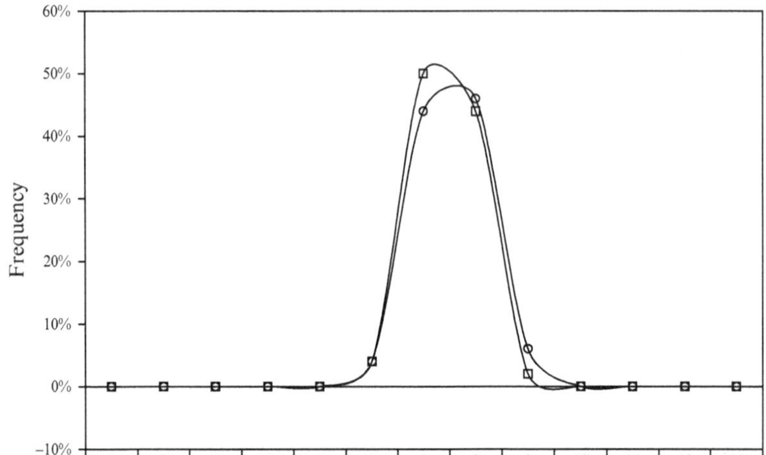

	1	2	3	4	5	6	7	8	9	10	11	12	13
Random	0%	0%	0%	0%	0%	4%	44%	46%	6%	0%	0%	0%	0%
Small-World	0%	0%	0%	0%	0%	4%	50%	44%	2%	0%	0%	0%	0%

Fig. 12.33 Comparison of frequency tables
Source: Author

the volatility of expected revenues and the respective cash flows. For this reason, the complex networks simulator can be applied in order to derive additional insights on the diffusion dynamics of complex customer networks. The presented approach can used to in order to approximate the path-dependent development of the underlying and of the volatility, which are vital input factors for valuation in software markets. The results are consistent with our findings from another research results. The distribution of customers of the small-world network is smaller, than the distribution of the random network model. Earlier investigations revealed that network topogies have an impact on the diffusion process. Another finding was that random networks percolate faster than small-world networks. All findings are confirmed. Moreover, the comparison reveals the systematic differential between both approaches. Hence, it is important to note that analytical valuations are likely to overvalue the customer network by ignoring the topological structure.

12.4.3 Reconsideration of Contributions to Customer Network-Centric Valuation

The previous research highlights the vital importance of network effects in complex customer networks. While classical research assumes implicitly random networks, our investigations indicate that customer networks tend to be small-world networks. The comparison reveals that the network topology of the customer network has an impact on the volatility of expected revenues and the respective cash flows. For this reason, the complex networks simulator can be applied in order to derive additional insights on the diffusion dynamics of complex customer networks. Simulations can used to approximate the path-dependent development of the underlying and of the volatility which, in turn, are vital input parameters for valuations. A prerequisite is, however, that the required information for the simulator is accessible and reliable. In summary, the previous analysis supports the hypothesis that the complex networks perspective on network effects in complex customer networks contributes enhance the quality of valuation in software markets.

12.5 Social vs. Natural Scientific Networks

In this final section of the complex networks analysis the social nature of customer networks is contrasted to natural scientific networks. Hence, the following investigations are pursued. In the next subsection, the hypothesis is derived and its underlying motivation is presented. In a second step, the hypothesis is challenged in an analysis based on complementary investigations and simulations. Finally, the most important findings are summarized in a reconsideration of the hypothesis.

12.5.1 Hypothesis on Social vs. Natural Scientific Networks

Although complex networks theory reveals a variety of universal properties and dynamics of networks, it is also necessary to be aware of the underlying assumptions and the respective limitations of this interdisciplinary research approach if it is applied to customer network-centric valuation in software markets. For this reason, we hypothesize that customer networks in software markets are special due to their social nature which is different from electronic, information or other scientific networks [HP5]. This hypothesis is motivated by shifting the focus of the research to the limitations of the interdisciplinary approach.

12.5.2 Analysis of Social vs. Natural Scientific Networks

There are a variety of supportive findings that social networks have considerable different properties than scientific networks. A comparison of complex social customer networks with other types of networks reveals their peculiar properties.

1. Complex Networks Metrics of Social Networks

The comparison of empirically investigated complex networks provides a variety of insights on the peculiar nature of social networks.[25] Social networks are rarely organized as random networks. In contrast to random networks which have a binomial or Poisson degree distribution, real-world networks tend to have a highly

[25] Please confer the overview on empirical complex networks studies.

right-skewed distribution of degrees. In random networks edge there is an equal probability for edges which yields a binomial or Poisson degree distribution in the limit of large networks (Erdos and Renyi 1959). Real-world networks, in contrast, have frequently a highly right-skewed distribution with a long right tail of values that is difficult to measure (Broder et al. 2000). Moreover, a comparison between real-world networks and random graphs with a similar number of vertices and edges reveals that the cluster coefficient tends to be considerably higher in real-world network structures (Albert 2001).

2. Small-World Properties of Social Networks

As previously stated the social customer networks tend to have a small-world topology. For this reason, the previously stated small-world studies indicate that most social networks are small-world networks also in this context.[26] Recent research investigating the structure of social networks reveals that the average path length of messenging networks, such as Microsoft Messenger, is 6.6 (Leskovec and Horvitz 2008). These findings are in line with similar studies on email networks (Dodds et al. 2003).

3. Delineation of Social Networks

Moreover, all model building activities have in common that the designer of a partial network analysis is confronted with the dilemma to determine the boundaries of social relations which are extremely difficult to delineate because of their fuzzy nature. Particularly, in partial network analysis, a legitimate network separation is ignored by some network researchers which leads to inaccurate results, since some emerging properties of the overall entity are not identified. Therefore, it is necessary to either make a theoretically informed decision about the significant boundaries of the model based on statistical theory or to work with the total population. However, real-life networks are generally very large, the complexity of the computational problems that need to be solved in the analysis yield considerable research obstacles that can only partly be overcome by numerical approximations simplifications, and simulations.

4. Network Resilience of Social Networks

All investigated social networks have in common that they are organized around some central nodes and their growth can be described by the outlined principle of preferencial attachment, i.e., the higher the degree of social node, the higher the chance that new links are connected. This has vital implications on their network resilience. It makes them particularly vulnerable to targeted attacks, e.g., the disintegration of the internal and external corporate network in a corporate turnaround (Hommel and Kemper 2006).

---

[26] Please confer Sect. 12.3.2.

12.5.3 *Reconsideration of Social vs. Natural Scientific Networks*

Although the outlined research revealed many analogies of social networks to scientific networks there are also some considerable differences. As these can influence significantly the properties and dynamics of the network, it is vital to account for such differences. Hence, the outlined findings support the hypothesis that there are some important differences between social and scientific or technical networks. Based on these findings, a context-specific analysis of complex customer networks is integrated in the complex networks framework for valuations in software markets. Consequently, it is important to be aware of the specific characteristics of social networks if the complex networks theory is applied to social customer networks, e.g., with respect to the stability of the links. The complex networks methodology provides only reliable insights, if the analysis accounts for the peculiar nature of social networks.

12.6 Reconsideration of the Complex Networks Analysis of Customer Networks

The complex networks analysis of customer networks revealed a variety of new insights on the previously developed hypotheses. While most hypotheses were confirmed, the hypothesis on the scaling properties of complex networks was rejected and modified. A summary of the insights is provided in the following.

1. *Diffusion dynamics in varying network topologies.* Our theoretical and numerical comparison of diffusion dynamics in various network topologies revealed, that the topology of a network has an impact on its dynamics. Therefore, it is important to consider the topology of customer networks in the analysis of network effects in software market valuations, e.g., with an implementation of a customer networks-centric valuation approach. Due to this insight, the previously developed framework for valuation in software markets is modified.
2. *Scaling properties of complex customer networks.* Alternative research suggests that the scale of the network does not have an impact on its properties and dynamics, but no proof for this hypothesis. As such scaling properties of complex networks are of primary importance for the research question, they were investigated with theoretical contributions and simulations. In essence, our analysis suggests to reject the general form of the hypothesis. The investigations revealed that the scaling properties of complex networks depend on the underlying network topology of the customer network. Thereby, the research closes an open research gap with a counter-intuitive finding. While the scale of the network is irrelevant in random customer networks, it can be relevant in customer networks with a small-world or a scale-free topology. Since previous investigations revealed that customer networks in software markets are primarily small-world

networks, the finding has to be considered in the design of the complex networks valuation framework for valuation in software markets.

3. *Network topologies of complex customer networks.* Our investigations revealed that customer networks in software markets are primarily small-world networks. Hence, it is important to consider the implications of the small-world customer network topology in the subsequent investigations and in respective software market valuations. Therefore, specific investigations that allow one to determine the exact network topology of complex customer networks should be integrated into the modified version of the valuation framework.
4. *Contributions of the complex networks analysis to customer network-centric valuation.* The network effects research emphasized the importance of network effects in customer networks. Our investigations revealed, that the complex networks adoption and diffusion simulator can be applied in order to derive insights on the diffusion dynamics of complex customer networks in software markets. In particular, it allows one to derive additional information into the underlying and on the volatility for valuations in software markets. This is another reason for integrating a complex network analysis to a valuation framework for valuation in software markets.
5. *Social vs. natural scientific networks.* The comparison of social networks with natural scientific networks revealed the peculiar properties of social networks. Consequently, it is even more important to be aware of the specific profile of customer networks. This is particularly crucial, if insights from complex networks theory should be applied to phenomena in socials. Special properties, e.g., the resilience of networks, can differ significantly. The insight is another reason to account for the peculiar nature of social networks in valuations in software markets.

The findings of our complex networks analysis of customer networks indicate that it is necessary to modify the previously developed network effects framework for valuation in software markets. Hence, the new insights are integrated in order to design a complex networks framework for valuations in software markets that accounts for properties, topologies, and dynamics of complex customer networks. These modifications are considered in the following chapter.

Chapter 13
Complex Networks Framework for Valuation in Software Markets

> "I simply wish that, in a matter which so closely concerns the wellbeing of the human race, no decision shall be made without all the knowledge which a little analysis and calculation can provide."
>
> (Bernoulli 1760)

The examination of the previously designed network effects framework revealed research opportunities by considering properties, dynamics and topologies of networks. The goal of this chapter is to close the identified research gap. Hence, the previously designed network effects framework for valuations in software markets is modified based on the derived insights of the complex networks analysis of customer networks. This modification is achieved in a two step process. First, an overview of the framework and the modifications based on the complex networks research is provided, before the individual phases are described in detail.

13.1 The Complex Networks Valuation Framework

Network effects are still considered to be highly relevant to software markets. Therefore, the previously developed network effects framework for software markets is the basis of the new concept. But the reconsideration of the network effects framework indicated its limitations with respect to network properties, topologies, and dynamics. For that reason, complex networks research was conducted. The literature review, the design of the simulator, and the complex networks analysis of customer networks of the previous chapter revealed a variety of insights that are incorporated in the new version. Consequently, a complex networks valuation framework is developed which consists of the following four interdependent phases that are adopted to the findings of the complex networks research.

1. Corporate and software market analysis
2. Complex networks software market model
3. Complex networks software company valuation model
4. Sensitivity analysis

In detail, the following modifications are conducted in the various phases of the framework.

A. Kemper, *Valuation of Network Effects in Software Markets*,
Contributions to Management Science, DOI 10.1007/978-3-7908-2367-7_13,

1. *Corporate and software market analysis.* In the first step the corporation and the software market are investigated. The purpose of this analysis is to develop a sound understanding of the business model for the subsequent complex networks software market model, which is different than in the network effects framework. In addition to the previous version, customer networks are investigated in order to develop a better understanding for the relevance of customer networks for the investigated company.
2. *Complex networks software market model.* In the second step, the software market model is modified. A valuable insight is that complex networks methodologies can be applied in order to analyze the network metrics and the topology of complex customer networks. Hence, the underlying customer network is investigated from a complex networks perspective. This complex networks analysis of the customer networks is integrated in the market modeling process into a complex networks software market model. The focus of this analysis is on the network topology of the customer network in order to design a suitable customer network model with the adoption and diffusion simulator based on relevant network theoretical metrices. It consists of an explicit analysis of the customer network based on the available data. Based on this data, the topology of the customer network is investigated and used as a basis for the design of complex networks software market model with the simulator. This model allows to investigate relevant network metrics of complex customer networks, e.g., the diameter or characteristic path length of the customer network, in order to develop a better understanding. In addition, adoption and diffusion dynamics of the complex customer network can be investigated. Such investigations increase the transparency with respect to the customer network, the relevant network effects and allow to derive probability distributions for the development of the revenues deriving from the customer network. These information are vital input parameters for the following software company valuation model.
3. *Software company valuation model.* The previously outlined complex networks software market model allows the modeling the price of the underlying and the volatility of the option on a finer level of detail. Instead of directly modeling the cash flows, they are decomposed into genuine revenues that result from sales in customer networks. Consequently, it is necessary to model the customer networks and to derive a reliable description of their development. In contrast to traditional Markov approximations of the underlying asset in which all states are assumed to be independent of prior states, previous findings suggest that the development of customer networks are interdependent. Hence, the developed adoption and diffusion simulator is integrated in order to provide more accurate descriptions of customer networks for valuations in software markets. Based on the complex networks insights of the outlined analysis, the optimal valuation model is chosen and implemented. Thereby, the model processes additional information on network properties and dynamics that contribute to an increasing quality of cash flow projections and in turn to better valuations.
4. *Sensitivity analysis.* Finally, a sensitivity analysis of the valuation model challenging the results concludes the framework. Again, it is important to note that it

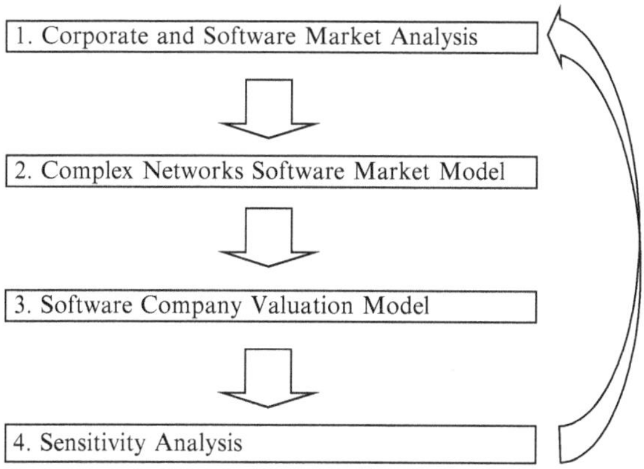

Fig. 13.1 Complex networks valuation framework
Source: Author

may be necessary to reiterate some steps of the framework. This can be necessary as new insights of one phase may have an impact on other phases. Particularly the calibration of the complex networks software market model may required some time and iterations before it delivers reliable data for the valuation model. The sensitivity analysis is intended to challenge the results in order to increase the reliability of the results.

The interdependencies of the various phases are summarized in the subsequent overview (see Fig. 13.1).

13.2 Corporate and Software Market Analysis

In the first step the valuation target and its software market environment are investigated. Hence, the business model of the valuation target is analyzed with a focus on its vital revenue sources, its value drivers, and its cost structure. This first phase of the framework is intended to provide an overview on the valuation target and its software market environment. For this purpose, the structure of the software market segment is investigated, before the most relevant value driver and the cost structure of the company are analyzed. The goal is, again, to determine the operating performance from an investors' perspective. As the total turnover of the company is a product of the quantity of units sold and the respective prices, it is necessary to identify the underlying drivers of the quantity of sales and the respective price level. All of these factors have to be assessed in order to understand the business model and to provide a platform for the following complex networks analysis of the complex customer networks.

13.2.1 Structure of Software Markets

Similar to the old version, it is necessary to develop a systematic approach to the market analysis, as different segments of software markets follow distinct rules.[1] In addition to the classical industry specific aspects of the industrial organizational structure-conduct-performance paradigm, it is necessary to consider also network effects due to their stated relevance for software markets.[2]

1. *Customers.* An analysis of the customers is central to all business models as they determine the demand for the product. Hence, it is important to understand the background, the market power and the cost-benefit considerations of customers. The previous analysis revealed that interdependent consumer decisions are frequently based on network effects.[3]
2. *Suppliers.* In classical industries the background, the market structure, and the incentive system of suppliers are important for the assessment of production restrictions. Software companies are rarely restricted by suppliers in the development, reproduction or distribution of software as these are frequently distributed via the Internet.[4]
3. *Competition.* According to the classical model, competition has a decisive impact on the supply and price strategy of companies. If the model is applied to software markets, it is important to consider the structure and the motivation of the direct and indirect competition as such aspects are likely to influence the competitive pressure as well. Actions and reactions of competitors may be integrated based on game-theoretical considerations, which in turn increases the complexity of investigations.
4. *Market entry barriers.* Market entry barriers represent potential changes in the competitive landscape and imply a game-theoretical uncertainty. If entry barriers are low and profits high, it is likely that other competitors will be attracted. In contrast, high market entry barriers such as large installed bases or high upfront investments, are likely to deter potential competitors.
5. *Substitutes.* Rivaling technologies which are perceived by customers as substitutes are relevant as they are likely to influence the revenue streams of the software companies. If such substitutes are sufficiently attractive, customers switch the product at some point. Hence, substitutes have a decisive impact on company operating in software market segments whose are strongly driven by network effects.
6. *Network effects.* As previously outlined, the underlying communication and interaction backbone, i.e., the customer network, has to be investigated in order to

[1] Please confer Sect. 8.2.1.

[2] Please confer (Porter 2008) for the updated five forces industry model and for further information on the structure-conduct-performance paradigm.

[3] Please confer Sect. 3.3.

[4] Please confer Sect. 3.3.8.

specify network effects. Possible are local interactions, i.e., word-of-mouth referrals, or implications from long-range interactions, i.e., media communication.
7. *Customer networks.* A new aspect is to compare the existing customer networks in the relevant software market segment. Accordingly, it is important to collect data on their size, their network characteristics and their network topologies. This is a preparation for the design of the complex networks software market model.

13.2.2 Value Driver of Software Companies

Similar to the old framework, the focus of the analysis is shifted to an internal company perspective after the previously investigated external market view. Based on a solid understanding of the software market segment, the main value driver of the company are considered. Depending on its maturity, the following main value driver of software companies can be identified (Wirtz and Kam 2001; von Westarp 2003; Bassen and Popovic 2004; Maaß and Pietsch 2008).[5]

Value Driver in the Innovation Phase of Software Companies

Companies in the innovation phase are startups with innovative products in the beginning of the company life cycle approach. Their financial performance is characterized by low revenues and losses. As they try to reach the critical mass in order to let their customers benefit from increasing network effects, the focus is on the customer network, the market position and the partner network. The value driver can be operationalized based on various indicators:

Customer network. The customer network can be described by its size, its network metrics, its network dynamics and its stability.
Market position. The market position is determined by the user intensity of the products and the market share.
Partner network. An external partner network is a considerable value driver as it allows to share marketing, development or sourcing activities.

Value Driver in the Expansion Phase of Software Companies

Companies in the expanion phase generate revenues and grow rapidly. Hence, revenues of the software company become increasingly important. In summary, the main value driver of this phase are the customer network, the market position, the

[5] Please note the similarity to the phases of the previously presented product life cycle in Sect. 3.3.1.

partner network and the turnover. These value drivers can be quantified with the following indicators.

Customer network. The customer network can be described by its size, its network metrics, its network dynamics and its stability.
Market position. The market position is determined by the user intensity of the products and the market share, e.g., in comparison to top competitors. As the costs to win a customer are six times as high a to retain a customer, the customer retention rate should be as high as possible. The customer churn rate should be as low as possible, in turn (Kalkota and Robinson 2001).
Partner network. An external partner network is a considerable value driver as it allows to share marketing, development or sourcing activities.
Financial key performance indicators. Financial key performance indicator such as revenues, contribution margin are becoming increasingly important in this phase.

Value Driver in the Maturity Phase of Software Companies

Mature software companies are established market player with stable cash flows generating profits, but with decreasing growth opportunities. Therefore, their growth rates decrease in comparison to the initial phases and to growing competitive start-ups. In this phase the classical value driver of company valuation are relevant. These value driver can also be quantified with the following indicators.

Customer network. The customer network can be described by its size, its network metrics, its network dynamics and its stability.
Market position. The market position is determined by the user intensity of the products and the market share, e.g., in comparison to top competitors.
Partner network. An external partner network is a considerable value driver as it allows to share marketing, development or sourcing activities.
Financial key performance indicators. Financial key performance indicator such as revenues, contribution margin are becoming increasingly important in this phase.

Due to their importance for valuation in software markets, particularly during the innovation phase of software companies, the revenues of the company are investigated in detail. A key to understand the revenues of software companies is the relationship between software prices and sales, i.e., the price-quantity relationship of the business model.[6] For this reason it is necessary to investigate the sales and the pricing of the software on a finer level of detail, e.g., in a commercial due diligence (Niederdrenk and Maack 2008).

[6] Please note that revenues are a product of sales volume times price.

Sales of the Software

The quantity of sales are influence by a variety of factors.[7] Previous research reveals that the following aspects are most relevant to software markets.

1. *Potential market.* Sales are determined by the maximum size of the market. This determines further growth and market opportunities as there is a limit to growth.
2. *Number of customers.* The number of actual customers determines the contemporary revenues. It is highly relevant as software companies need a critical mass of customers for a sustainable business model that is successful in software markets.
3. *Word-Of-mouth marketing.* The research on network effects reveals that software is frequently bought due to direct recommendations of the social environment, e.g., friends, colleagues or family. Hence, customer communication by word-of-mouth marketing is a decisive factor effecting sales that has to be evaluated individually.
4. *Product features.* The product features are decisive factors for the purchase of software in software markets. Although programs tend to provide more functions than needed by an average user, e.g., Microsoft Excel or Word, users frequently require a minimum set of core functionalities.
5. *Structure of customer networks.* The previous research emphasized the importance of customer networks. Hence, it is necessary to investigate the referral revenues, the number of opinion leader in the customer network and further indicators such as the churn rate. In this context it is important to note that the customer retention rate should be as high as possible, as the costs to win a customer are six times as high a to retain a customer (Kalkota and Robinson 2001). The customer churn rate should be as low as possible, in turn.

This selection of parameters is not exhaustive, and specific to the respective software market segment, but the outlined factors are highly relevant to valuation in software market and should therefore be incorporated in the design of a network effects software market model.

Price of the Software

Similarly, the price-quantity relationship is determined by the price of the software. This, in turn, is controlled by supply and demand. While the supply of software is virtually not restricted due to its availability via the Internet, the demand for software can be determined based on the following two most important factors in software markets:

1. *Generic utility of the software.* In general, the willingness-to-pay of a customer is related to the perceived usefullness of the product. Hence, the generic utility of the software is a driver of the price.

[7] Please confer (Meffert 2000) for a broader overview.

2. *Network utility of the software.* As network effects determine the benefit of software, they influence also the among that a customer is willing to pay for it. The influence increases with a growing customer network.

According to empirical studies on the pricing of software, this list is not exhaustive but explains a significant percentage of the pricing mechanisms based on hedonic pricing models.[8] Hence, these factors should be included in a software market model.

13.2.3 Cost Structure Analysis

In addition to the analysis of the topline, i.e., the revenues, it is also necessary to investigate the cost structure in order to determine the operative performance of software companies. The following cost factors should be taken into consideration in a software market model.

1. *IT costs.* Research reveals that software companies are frequently subject to a combination of high fix costs and marginal variable costs (von Westarp 2003). This constellation reinforces the emergence of network effects. The costs result mainly from the setup of a technological infrastructure and for required software licences.
2. *Labor costs.* Labor is a vital cost factor for software companies as esearch, product development, and testing are very labor-intensive. The software has to be specified, developed and documented. This is done by well-paid software engineers who are either employees or freelancer.
3. *Marketing costs.* Depending of the software market segment, the marketing costs for software can be marginal or gigantic. The increasing number of software companies induced increasing marketing costs over the last years, particularly if a company is new and unknown. As the costs to win a customer are six times as high in comparison to the costs for retaining a customer, the customer retention rate should be as high as possible (Kalkota and Robinson 2001). The customer churn rate, in turn, should be as low as possible.
4. *General costs and administration.* Another considerable cost driver of software companies are the rents, overheads and other fixed costs, e.g., office equipment.
5. *Expansion costs.* During the expansion phase the costs can increase rapidly. While *modernization costs* are required to update the product functionality due to a relatively short lifecycle, it is also necessary to consider the costs for *extensions* of the customer network. Particularly, the costs for an adequate IT infrastructure and for labor can increase exponentially. Moreover, it is necessary to consider maintenance and update costs which can account for approximately 70% of the total software costs during the total product development cycle (Berger 2008). In addition, the increasing customer base may require technical upgrades in order

[8] Please confer Sect. 5.3 and (Groehn 1997) for further details.

to account for the larger number of customers, e.g., due to the necessity to open a call center or to hire external agencies and consultants.

Based on the value drivers and the respective cost structure of the individual software company, a software market model is designed in the following section.

13.3 Complex Networks Software Market Model

In this section the design and analysis of a complex networks software market model is described based on the outlined complex networks adoption and diffusion simulator. This step has the most significant modifications with respect to the network effects framework. The goal of this phase is to gain additional insights into the value of the underlying and on the volatility which, in turn are vital input parameters of the valuation. Research comparing the similarities and differences among social and natural scientific networks revealed that it is important to conduct a context-specific analysis.[9] This research insight is incorporated at this stage in the analysis of the complex customer networks in software markets. Accordingly, a systematic complex networks software market model comprises the following interdependent phases.

1. Scale and scope of the complex networks software market model
2. Topology of the complex customer network
3. Implementation of the complex networks software market model
4. Simulation of the software market development
5. Derivation of complex networks data for valuations in software markets

13.3.1 Scale and Scope of the Complex Networks Software Market Model

Before the actual model is designed, it is necessary to determine its scale and scope. Depending on the optimal scale of representation, the size of the network can vary from relatively small, e.g., some individuals, to very large, e.g., groups, populations. The following parameters determine the optimal scale and scope of the complex networks software market adoption model.

Network size. The size of the population is equivalent to the number of interacting agents. As the complexity of numerical software market models increases exponentially, i.e., over-proportionally, with an increasing size of the network it is important to be aware of the complexity trade-off. The larger the simulated network, the higher is the level of realism but the higher are also the computational requirements. In general, the marginal contribution of an additional level

[9] Please confer Sect. 12.5.

of detail decreases with increasing level of detail. Hence, it is possible to compare the advantages and costs of various levels of representation in order to determine the most reasonable scale for the investigated software company.

Connections between customers. As research on social networks reveals that social interactions in a customer network depend on their relationship to each other, the number and type of links between customers determines the level of detail of the software market model. The more complex the relationship the higher are computational requirements. Prior research on social networks reveals that bidirectional links representing social relationships are a reasonable concept to model interactions among customers.

Benefits of the software. According to the decentralized standardization model the incentive to purchase a software is the sum of a generic and of a derivative utility. While the generic utility is derived from the product itself, the additional derivative utility accounts for the identified network effects. Depending on the adoption rule, the description of the utility can influence decisively the scale of the model.

Costs of the software. It is also necessary to quantify the opportunity costs for purchasing the software. Such costs result from the initial purchase or from additional maintenance or update costs. Similar to the modeling of the benefits, the level of detail of the adoption rule and the respective opportunity costs determine the scale of the model.

13.3.2 Topology of the Complex Customer Networks

Previous complex networks findings underline the vital importance of network topologies and their impact on dynamics in networks. Hence, it is necessary to determine the topology of the underlying customer network and to conduct further investigations that allow one to simulate the customer network in the complex networks adoption and diffusion simulator. For this purpose, data on the complex customer networks is required. Empirical investigations of the average connectivity and the distribution of links in the customer network provide suitable insights.[10] Previous research revealed that there are a variety of network topologies, but that customer networks in software markets tend to be small-world networks.[11] The following structured analysis allows us to investigate the topology of the relevant customer network.

1. *Empirical survey.* In a first step empirical data on the network is required.
2. *Delineation of relevant Network.* Based on the data it is possible to determine the scale and scope of the investigated network. In this context, it is important to note the difficulty to investigate partial networks.

[10] Please refer to Sects. 10.3 and 12.2.2.

[11] Please refer Sects. 10.3 and 12.3.

3. *Observe network metrics.* Once the relevant network is delineated, relevant complex networks metrices, e.g., the link and node distribution, the diameter and the clustering of the customer network are studied.
4. *Determine topology of network.* If all relevant metrices are available, the topology of the network can be determined, e.g., the combination of a high clustering coefficient and a low characteristic path length suggest that the customer network is a small-world network.
5. *Generation of complex customer network model.* Once the relevant topology is identified, the complex networks adoption and diffusion simulator can be used to generate a network model with similar topological characteristics as depicted in the following subsection. Depending on the required accuracy, the calibration of the simulator may require some time for the fine tuning of the model.

13.3.3 Implementation of the Complex Networks Software Market Model

If the generated network instance is representative for the investigated complex customer network, network diffusions can be simulated. In this simulation process it is crucial that the implemented adoption rule represents the adoption decision of real world customers. An overview of the implementation is provided in the next subsection. It is similar but slightly more complex than the network effects framework. This is a conceptual overview that is refined in the upcoming chapters of this book.

1. Generation of a customer network instance
2. Selection of a software adoption algorithm
3. Storage of the simulation data
4. Analysis of the network data
5. Representation of the network and results

The five phases of the implementation are described in detail in the following subsections.

Generation of a Customer Network Instance

In the first step of the implementation phase, it is necessary to generate a representative network instance, i.e., a network representation, in which the product adoption and diffusion process is simulated. For this purpose, it is assumed that each product has an installed base consisting of n independent actors within a population of p network participants who are connected by communication ties, representing friendships, family, or business relationships. The goal of this phase is to create a network instance with representative network properties.[12]

Selection of a Customer Adoption Rule

In the second step, the adoption decision of customers is codified based on the identified relevant decision parameters. This selection of a representative adoption algorithm is a crucial issue of the model that determines the reliability and the quality of the model. As there are a variety of approaches to model adoption decisions, it is necessary to select the most suitable approach based on the outlined relevant decision variables. Again, a modified version of the numerical decentralized standardization model is applied, as it accounts for network effects also from a complex customer network perspective.[13] Based on this decision the following refinements of the adoption rule have to be specified.

Set-up costs. Set-up costs are the initial purchasing costs. If competing products are considered, a switch to other software products may require the payment of set-up costs more than once. It is important to note that initial costs are sunk costs in following phases.

Maintenance costs and subscription fees. In addition to the initial set-up costs, some software products are subject to additional subscription or maintenance fees.

Autarky utility. Previous research revealed that a software can provide an autarky benefit. This is independent of other users as it describes the generic benefit of the software, e.g., a tax calculation program or a vocabulary training program.

Direct network effects. Direct network benefits result from an expanding customer network. This allows to benefit from each neighbor directly by increasing possibilities to apply the software with network members. In turn, direct network effects are frequently also related to additional costs per link, e.g., time to add or maintain a link.

Indirect network effects. Indirect network effects occur also from an expansion of the customer network, but they are generated from indirect, i.e., not directly connected people of the customer network. They can also be related to additional indirect costs.

Net benefit. The net benefit is the residual of the positive sum of the benefits deriving from autarky, direct network and indirect network effects minus the respective costs. Based on such decision parameters a net benefit coefficient can be computed. Please note that this cannot be negative as the software would not be purchased.

---

[12] Please note that the research on Complex Networks distinguishes a variety of network typologies which are discussed for didactic reasons in the next part of the book. Please confer IV.

[13] Please consider Sect. 7.2.

Estimation of adoption probability. Each agent has to form an opinion concerning the adoption probability of other users. There are a variety of approaches for resolving this game-theoretical problem.[14]

Duration of adoption process. The duration of the adoption process is a factor of primary interest. While in some software markets an equilibrium is reached after few iterations, other may require many iterations. Some will never reach a stable equilibrium. If the model reaches a steady state it is important to investigate the correlation of the network effects and the time to the steady state of the model. In this context it is possible to analyze the shortest diffusion path, which is the minimum amount of time that is required until a stable state equilibrium emerges in a given customer network. In other words, this can be the minimum time period that the company has to finance its operations in order to reach a critical mass of customers in the special case that the equilibrium is the critical mass and the net present value is positive.

Based on the outlined assumptions, the software market model allows to approximate a distribution of revenues which is required in order to derive a reasonable distribution of cash flows for the software company, which, in turn, determines the valuations in software markets.

Storage of the Simulation Data

In the third implementation step, the results of the simulation have to be stored for subsequent analyses. Once the software market model is developed, it is iterated in order to simulate a variety of possible developments. There are several options to export the generated data, but since the data volume is significant in large customer networks it is important to implement efficient solutions. This comprises also an efficient interface for storing the data in a database for further investigations.

Analysis of the Network Data

The fourth step of the implementation is the analysis of the generated complex networks data. This core element should provide a network theoretical toolkit for a static and dynamic network analysis. Hence, network theoretical tools are applied in order to derive network characteristics and to identify network dynamics of the analyzed software segment. Depending on the desired level of analysis and the required operations, the analysis provides more or less complex functionalities. A vital information is the number of adopters after a certain number of iterations.

[14] Please confer Chap. 7 for a discussion of various adoption approaches.

Representation of the Network and Results

Finally, the network and the generated data are reported to the user. Depending on the size of the complex customer network, a variety of visualization options are available. In general, the goal is to provide in addition to numbers also visualization routines that depict the network, the adoption of the software and the findings of the network analysis, e.g., with Bayesian networks. Such a visualization of the software adoption and diffusion process allows an intuitive representation of the results and is also a plausibility check.

13.3.4 Simulation of the Software Market Development

After the design of the complex networks software market model, multiple simulations should be performed in order to investigate the probability distribution of various outcomes. Each run represents a possible market development. Many paths are simulated in order to derive a representative probability distribution for the respective parameters. The results of the multiple runs are aggregated and interpreted with respect to their frequency. Such a procedure allows one to determine the probability distribution for various results. A variety of additional aspects can be relevant to the investigated research questions.

Individual adoption and diffusion paths. Single adoption and diffusion paths can be depicted in detail. This allows to identify the relationship of different network effects to each other. For example, it is possible to derive valuable insights on the process such as that the initial phase of a software adoption and diffusion process is rather more determined by the autarky utility, whereas network effects tend to become more important in a later phase if the number of customers increases.

Relationship of network effects. As network effects can result from various sources, the relationship and proportion of all network effects is another interesting research focus. Accordingly, it is possible to outline the development of the indirect network effects in comparison to the direct network effects. Such insights, their relative strengths and their development can be particularly relevant to valuations as they reveal insights on this potential value driver.

Total net benefit per capita. Depending on the assumptions concerning the expected direct and indirect network utility, the total net benefit of the product can be computed per individual.

Once the simulations are conducted, the relevant data is derived for the valuation.

13.3.5 Derivation of Complex Network Data for Valuations in Software Markets

In a final step of the implementation, the relevant complex networks data is collected for valuation purposes. This is also similar to the previous network effects version, but in addition further information on the properties, the topology, and on the dynamics of the customer network can be taken into consideration. Depending on the quality of the network data, the software market model allows to enhance the quality of the valuation. In this context the following parameters are particularly relevant to valuations in software markets:

1. *Number of adopters.* The number of adopters allows one to describe a range of software users over time. This is particularly important if the goal of the analysis is to derive an approximation of the financial structure and performance at a specific point in time, e.g., after 1 year. Multiple simulations and sensitivity analyses allow to derive insights on the customer network size for various simulation parameter configurations.
2. *Revenues.* Based on the approximated number of customers, it is possible to derive a range for the respective turnover of the software company, if the prices of the software can be approximated.
3. *Cash flows.* If relevant costs and depreciations can be approximated for the projection period, the outlined approach based on the software market model allows to determine a range for the expected cash flows.
4. *Shortest diffusion path.* The shortest diffusion path represents the minimum required time until a stable equilibrium emerges. This is equal to the minimum time period that the company has to finance its operations in order to reach a critical mass of customers.
5. *Analysis of the critical mass.* The critical mass of the customer network can be investigated and quantified by observing the size of the giant component. For this purpose the size of the giant component is investigated for various parameter combinations and based on multiple iterations. A diagram comparing the size of the results will reveal whether the customer network is subject to a phase transition, i.e., if there exists a critical mass of customers above which the diffusion process reinforces itself. Once the critical mass is identified, it is necessary to determine the probability for crossing the critical mass of customers based on multiple iterations.
6. *Identification of opinion leader.* Opinion leader can be identified by investigating the nodes with the highest degrees. If the degree distribution of all nodes is compared, it is possible to determine the number of nodes with a significant influence in the customer network. It is necessary to quantify the number of such opinion leader for the given customer network and to find them in real world markets based on network theoretical hub identification strategies, e.g., by personal referrals.

All in all, the outlined complex networks version accounts for additional network characteristics and dynamics of customer networks which are relevant for the valuation model, as depicted in the following section.

13.4 Complex Networks Software Company Valuation Model

The generated complex networks software market model provides vital information for the subsequent valuation model. Again, the total value of the software company is split into a passive stand-alone value and an additional real option value. The passive value is the net present value, which can be frequently derived with a DCF approach. In contrast, the additional optional value represents the value of managerial flexibilities, such as network effects in customer networks. Such flexibilities are valued as options based on their underlying cash flow distributions which can be approximated with the complex networks software market model. In essence, the complex networks software company valuation model is at this point a modified version of the classical real option valuation process and of the customer networks framework. Both consist of four interdependent phases:[15]

1. *Identification of main sources of uncertainty.* In the first step, the real option approach is framed. For this purpose asymmetric payoff structures are identified and indispensable characteristics of real options, such as flexibility, uncertainty and irreversibility, are investigated. Then, the scale and scope of the valuation model is determined based on the most significant sources of uncertainty.
2. *Selection of option pricing model.* Once all relevant real options are identified, the most suitable valuation approach, a simulation, is identified and the corresponding option pricing model is selected.
3. *Determination of the valuation parameters.* In the next step, the respective parameters of the selected option pricing model are determined with the simulation.
4. *Calculation of option values.* In a last step, the values of the options are determined and their interactions are considered. As the marginal contribution of additional options decreases, it is reasonable to concentrate on the most relevant options.

The outlined phases of the complex networks software company valuation model are summarized in the following sections. Based on more detailed complex networks information on the customer networks, they are similar but not identical to the previously outlined network effects framework.

13.4.1 Identification of Main Sources of Uncertainty

In the initial phase of the complex networks software company valuation model, the main managerial flexibilities of software companies, such as complex customer networks, are identified. They are framed as options based on cost-benefit considerations. For this purpose, the respective volatilities are approximated with respect to

[15] Please confer Sect. 2.3 for further details.

the uncertainty of its underlying. As in the previously developed network effects version, deferral options, scaling options, and liquidation options are the most relevant options in software markets.[16] But based on the complex networks data additional information it available to assess customer network specific uncertainties, the resilience or the diffusion dynamics of complex customer networks.

13.4.2 Selection of Option Pricing Model

In the second phase of the complex networks software company valuation model, the relevant option pricing model is chosen for the respective real option. The previously outlined discussion revealed that numerical simulations based on the decentralized standardization approach are suitable for most valuations in software markets as they allow to value even complex path-dependent payoff structures which result from the individual software adoption perspective.[17] Hence, the simulation of the complex networks software market can be applied for most valuations in software markets. Supportive complex networks investigations of the customer network provide additional insights on the relevance of network effects. The higher the relevance, the less reasonable is the application of analytical option pricing models, as these aggregate all data on the corporate structure by assuming a random network topology, which is uncommon in social customer networks of software markets.

13.4.3 Determination of Valuation Parameters

In the third phase of the complex networks software company valuation model, the parameters of the selected option valuation approach are specified with respect to the investigated software company. As previously outlined, the value of options is determined by the same six parameters as in the network effects valuation framework.[18] While some of them can be easily derived, others are more difficult to obtain (Hommel and Pritsch 1999). It is important to note that the previous complex networks analysis of the customer networks provides a variety of useful information on the required parameters. In particular, it extracts information from the customer network that can be used in order to derive the value of the underlying and the volatility of the option, if the required network information is available. Once all required parameters are derived or approximated, the value of the options can be computed.

---

[16] Please refer to Sect. 2.3.3 for the real options typology.

[17] Please confer Sect. 4.3.

[18] Please confer Sects. 4.4 and 8.4.3 for an overview on the parameters.

13.4.4 Calculation of Option Values

In the final step of the valuation model, the value of the real option is computed based on the previously derived information. If all option pricing parameters are available, the value of the option can be determined, again, with modified versions of existing option valuation approaches, such as the Schwartz and Moon model (Schwartz and Moon 2000). In the Schwartz and Moon model the volatility of the option is approximated as the expected cash flow distribution of stochastic simulations. This transfers the problem to the identification of a suitable stochastic process that describes the development of the underlying. Here, the complex networks software market model provides the information on the distribution and adds topological information that enhance the description of the expected customer network in comparison to a description by a Markov process. The reason is that the development in the customer networks are interdependent due to the network effects which has an effect on the respective revenues, the respective cash flows and therefore also on the valuation. Hence, the developed software market model based on adoption and diffusion models can be used in order to understand the market dynamics including the network effects, and also to find a suitable description for the underlying and its volatility. Due to the outlined importance of network effects, this customer network-centric valuation in software markets is superior to a purely standard stochastic processes, since the complex networks software market model accounts for network effects. In summary, the results of the previously conducted complex networks software market model are integrated into the complex networks software company valuation model in order to compute an enhanced customer network-centric option value of software companies.

13.5 Sensitivity Analysis

Similar as in the network effects version, a sensitivity analysis is conducted in the final step of the complex networks valuation framework. The goal of this sensitivity analysis is to challenge the findings of the complex networks market and valuation model. Hence, the impact of one or multiple parameters on the results of the model are investigated in order to explore its stability and the most relevant value drivers. In addition, the profile of the software market revealed that the segments have individual characteristics which have to be taken into consideration in the valuation of software companies. Hence, the results of the sensitivity analysis are cross-checked with respect to the characteristics of the individual software market segment. Again, further important issues of a sensitivity analysis are the first-order or second-order interactions among multiple real options.[19]

---

[19] Please confer Sect. 8.5.

In the next chapter, the advantages and limitations of the complex networks framework for valuation in software markets are reconsidered. Although the new version accounts for complex networks phenomena, there exist still a variety of further research opportunities. These are outlined in the following final part of the book after the reconsideration of the complex networks framework.

Chapter 14
Reconsideration of Complex Customer Networks in Software Markets

"From a cocktail party to a terrorist cell, from an ancient bacteria to an international conglomerate - all are networks, and all are part of a surprising scientific revolution!"

(Barabasi 2002)

The pursued complex networks approach provides a variety of additional insights to valuation in software markets. A vital contribution of this research perspective is to underline the existence of various network topologies and network dynamics, which significantly influences the adoption and diffusion processes of customer networks in software markets. This implies that the topology of customer networks has a decisive impact on the development of sales in general and on valuations of software companies in particular. Traditional valuation methods, in turn, frequently implicitly assume that customer networks have unrealistic homogeneous random network topologies. This can result in significant distortions. Particularly, if strong network effects are present. Instead, it is necessary to investigate the properties of the customer networks with the developed complex networks adoption and diffusion simulator. For this reason, the relevant complex networks insights were integrated into a complex networks framework for valuations in software markets. This modification allows one to investigate the properties and dynamics of large-scale customer networks in detail. Despite such contributions, it is also important to note the limitations of the approach. These are comprised of, but are not limited, to the following three aspects.

1. *Utility Description.* A crucial issue in the design of the software market model is the quantification of the benefits of existing and potential consumers. While it is possible to quantify exactly the benefits of cost savings, e.g. cost reductions due to EDI, the description and quantification of benefits is a considerable problem in economic research in general. Depending on the existing market information and possible time constraints, additional marketing investigations are required in order to shed light on these important parameters. This, however, is beyond the scope of this book.
2. *Empirical Research.* The outlined results are subject to the assumed degree distribution. Therefore, further empirical investigations are required in order to derive more realistic distributions based on internal company data. An investigation of networks based on surveys is a challenging task for a variety of reasons. First,

A. Kemper, *Valuation of Network Effects in Software Markets*,
Contributions to Management Science, DOI 10.1007/978-3-7908-2367-7_14,

the individual perception of people is different. Particularly with respect to the strengths of a social network tie, opinions can be different. For example, if someone considers himself to be a friend of somebody else, this is not necessarily a reciprocal view. Our investigations reveal that most social complex networks are characterized by a few agents with a high connectivity while there are many actors with a relatively low number of contacts. With respect to software markets such distributions could be investigated by analyzing the structure of an Internet forum of a specific software. Such investigations provide insights on the topology and dynamics of a user community. Consequently, it is necessary to conduct market specific research in order to determine the relationship distribution of the customer in the investigated software market. Similarly, market network data can have a significant impact on the subsequent valuation. Hence, it is reasonable to challenge the underlying assumptions based on more specific empirical investigations. Reasonable approaches in this context are indepth case studies and large-sample studies.

3. *Network Generation.* As network topologies have vital implications on the diffusion behavior of networks, the accurate generation of networks is a decisive assumption of this approach. The results of a complex networks adoption and diffusion analysis are only a contribution, if the customer network is represented accurately by the generated model. For this reason, it is important to investigate systematically the available algorithms for network generation in order to develop reliable guidelines for network breeding in order to maximise the similarity between simulated and real world networks.

In summary, the complex networks framework exhibits a considerable potential for customer network-centric valuation in software markets. The outlined limitations, however, indicate that additional research is required. Therefore, some prospective research venues are presented in the final part of the book.[1]

[1] Please confer Chap. 18.

Part V
Summary, Implications, Limitations, and Outlook

The final part of this book provides an interdisciplinary review of the main findings, before target group specific implications are derived. Then, the limitations of the outlined research are discussed and prospective research opportunities are presented. The executive summary concludes this part.

Chapter 15
Summary of Findings

"In the longer run, network thinking will become essential to all branches of science as we struggle to interpret the data pouring in from neurobiology, genomics, ecology, finance and the World-Wide Web."

(Strogatz 2001)

In the first chapter of the final part, the most relevant findings are reviewed. First, general main insights are summarized. Then the results of the individual parts are recapitulated. Hence, insights on valuation in modern software markets, network effects in software markets, and customer networks from a complex networks perspective are reviewed, before the implications of these investigations are depicted in the following chapter.

15.1 Main Findings

Market-related issues of valuation are gaining increasing attention as managers and at the same time financial markets are also calling for an accurate valuation in software markets. The overall purpose of this research was to improve the quality of valuations in software markets based on innovative insights from network economics and complex networks research. For this purpose, we suggested a complex networks approach from a customer network-centric perspective. This accounts for network effects, network topologies and network dynamics in order to enhance the understanding of customer networks as valuable flexibilities.

Reasoning of Research

In essence, the research is based on the following reasoning. Customer networks are main value drivers of software companies and network effects are main drivers of customer networks. Hence, it is necessary to increase the understanding of customer networks in order to enhance the understanding of valuation in software markets. In the second step, it is important to note that customer networks are determined by the properties, topology and dynamics of the customer network, which can be

A. Kemper, *Valuation of Network Effects in Software Markets*,
Contributions to Management Science, DOI 10.1007/978-3-7908-2367-7_15,

investigated from a complex networks perspective. Hence, a complex networks perspective is applied in order to derive additional insights on customer networks for a better understanding of network effects which, in turn, are important value drivers of software companies that allow one to enhance the quality of valuations in software markets. This approach is summarized in detail in the following paragraphs.

Network Effects and Network Economics

In the first step, the profile of software markets revealed that the key to the valuation of software companies is customer networks. But despite this importance, it is not very popular in management theory nor in practice to account for the value of network effects in customer networks. An explanation is the peculiar nonlinear nature of customer networks, and their dynamics. Therefore, the we assumed a more market-oriented valuation approach by merging valuation, network effects and complex networks into a complex networks framework for valuation in software markets. While traditional research considers cash flows at a highly aggregated level, network effects are identified as decisive determinants of valuations in software markets and valued based on a real option approach. Hence, we disaggregated cash flows on the level of individuals in customer networks.

Complex Networks Perspective

The decomposition of the cash flows required a reliable description of the development of customer networks. This description was achieved with tools of complex networks theory. In summary, we developed a customer network-centric approach for valuation in software markets and the investigations underlined the importance of network characteristics and network topologies of customer networks for valuation in software markets.

Complex Networks Analysis of Customer Networks

Our research program revealed that the distribution of cash flows of an investigated software company can be approximated by a software market simulation based on a network adoption and diffusion model. But the simulator allowed further investigations of relevant hypotheses. Accordingly, the variation of customer network topologies has an impact on the dynamics of the customer network [HP1]. Moreover, our investigations revealed that customer networks in software markets are primarily small-world networks [HP2]. Insights on the scaling invariance of customer networks depicted a more diverse picture regarding the respective network topology [HP3]. Accordingly, up- and down-scaling is possible in random networks, whereas small-world and scale-free networks are not invariant to scaling. Furthermore, the investigations revealed that the outlined approach provides vital contributions to customer network-centric valuation in software markets [HP4]. Finally, additional research underlined that it is important to account for the special nature of social networks as they are different to natural scientific networks, e.g. since social links tend to be more fragile [HP5].

Applications of Framework

The complex networks framework has a variety of applications if the required information is available. Particularly, applications for turnaround assessements provide additional insights as a company with stand-alone negative DCF value can be a viable turnaround investment if the optional value due to the growth option of the installed base is considered. If the required funding for crossing the critical mass is considerably small in comparison to the potential returns, a turnaround investment can be financially profitable, despite a temporary negative net present value. In particular, as many software companies compete for markets rather than in markets, investments can be profitable despite large upfront investments. Hence, the costs and efforts required for the complex networks analysis can be a lucrative investment as it allows to increase the transparency in turnaround decisions. Due to the required efforts, it is necessary to decide on a case by case situation if the benefits of the additional of additional information weigh out the respective costs of a complex networks analysis.

15.2 Insights on Valuation in Modern Software Markets

Network effects are important phenomena in software markets, but despite their importance they are poorly understand. A review of existing traditional valuation approaches reveals several deficiencies. In particular, traditional approaches are not capable of accounting for managerial flexibilities. Hence, investment strategies often focus on a short-term perspective which ignores, in particular shortly before reaching a critical mass, the valuable managerial flexibilities of direct and indirect network effects. The consequence of this ignorance is suboptimal underinvestment strategies. Therefore, traditional approaches are reconsidered and requirements for valuations in software markets are derived. An assessment of the real option approach with respect to the requirements reveals that this approach has the potential to increase the quality of valuation in modern software markets. Within real option valuation there are a variety of valuation approaches to value the option such as analytical, quasi-analytical and various numerical approaches. A comparison of the valuation tools revealed that numerical simulations, based on software market models, are particularly well suited for valuations in software markets. But the required specification of the dynamics of the underlying and of the volatility are two challenging implementation problems. Traditional financial research frequently applies a standard normal distribution. While this assumption may be reasonable for some industries, e.g. crude-oil and natural resources, it is not reasonable in very volatile industries, e.g. software markets, which are governed by fat tails. Therefore, it is necessary to be careful in the specification of the distribution of the underlying asset as the distributions can differ significantly from the standard distribution. Because of the network effects, this recommendation is particularly relevant to valuation in software markets. While frequently the need for simplicity and applicability is a good advisor, errors can also provide misleading results if the underlying assumptions are over-simplified. As the trade-off between applicability and precision is particularly important in software markets, the analysis motivates the development of more sophisticated software market models in order to determine the respective input parameters for the valuation model. In summary, a real option valuation approach is suggested for valuation in software markets that is based on complex networks simulations of software markets.

15.3 Insights on Modeling Network Effects in Software Markets

Next, the identified research gap is investigated from a real option valuation perspective based on simulations of software markets that account for network effects by applying network adoption and diffusion models. A review of the relevant network effect theory, with respect to software markets, emphasizes the vital importance of network effects. Adoption and diffusion models allow the integration of network effects into a coherent framework for valuations in software markets. A review of the relevant literature reveals that the decentralized standardization model is a reasonable approach to model network effects in software markets. An integration of this approach in an adjusted software market simulation allows one to more accurately derive probability distributions as customer networks are interpreted as underlying assets. Research confirms the results, and indicates that the topology of a customer network is a decisive determinant of the software market model which is frequently neglected.[1] Therefore, the complex networks properties of customer networks are investigated in finer detail. All in all, a real option approach is suggested for valuation in software markets that is based on numerical simulations of software market models, and that accounts for network effects by applying a modified standardization model. The reconsideration of the derived insights illustrated that more detailed investigations of network topologies and dynamics are required. The profiling of customer networks in software markets demonstrates that their dynamics are not smooth and linear, but rather disruptive and nonlinear, since network

[1] For didactic reasons the structure of the book was chosen to accentuate this importance of network topologies and dynamics: First, the basic random network model was designed illustrating its defficiencies with respect to network topologies, before the complex networks perspective was adopted in order to investigate network properties and dynamics. This outlines the importance of network effects in a first step, but reveals also the limitations of network economics.

externalities are vital sources of risk and value penetrating all areas of management. The respective financing decisions can be rationalized, only if a critical mass required for financially stable business transition is quantified, since investments in such prospective projects can be interpreted as real options containing valuable flexibility. Thereby, the research illustrates the necessity to integrate network topologies and dynamics into corporate financial decision-making. This is approached from a complex systems perspective.

15.4 Insights on Modeling Customer Networks from a Complex Networks Perspective

The final part was dedicated to extend the discussion of valuation in software markets to a complex networks perspective. For this purpose, the complex networks literature is reviewed with a specific focus to contributions on properties, topologies and dynamics of networks. A multitude of alternative network typologies are identified, described and interpreted with respect to their implications on valuations in software markets. Since the topology of the customer network is identified as a vital determinant of the product diffusion process, a complex networks framework for valuations in software markets is proposed that accounts for network effects and, at the same time, also for varying network typologies. A comparison of this approach with conventional exclusively stochastic simulations emphasizes that the designed model provides an information theoretical advantage due to additional structural information. It can be extracted by a network analysis of the customer network, thus enhancing the specification of the cash flow distribution and, thereby, the quality of valuations in software markets. Network diffusion processes can be modelled analytically based on differential equations. A comparison with the numerical approach provides a variety of additional insights. Analytical approaches assume that customer networks are homogeneous. This implies the underlying assumption that customer networks have a random topology, whereas research revealed that software markets tend to have a small-world topology. Hence, analyical approaches are not suitable for software market models. The numerical complex networks adoption and diffusion simulator, in turn, resolves this problem as it is capable of implementing different network topologies, as well as varying diffusion processes for customer network-centric valuation. Although the design and implementation of such a complex networks simulator is difficult and requires an investment the resulting benefits in the form of additional insights for the valuation can be worth the additional costs. In each case it is necessary to evaluate the tradeoff between the benefit of more precise investment decisions and the required costs. However, it is important to note that the goal of the research is not to suggest that perfect forcasting of real world market developments. According to quantum mechanics such a Newtonian description of reality is impossible due to quantum phenomena on atomic scales (Shankar 1994). Nevertheless, the outlined concepts derive valuable insights as they are based on the underlying universal laws of complex networks which at least a "second best" solution that allows to enhance the quality of valuations at the expense of the required resarch which can be worth the effort under the outlined specific circumstances.

15.5 Reconsideration of Findings

The summary of findings reveals that the initial research objectives are achieved.[2] The *overall research goal* was to increase the quality of valuations in software markets by adopting a complex network perspective that allows the identification, quantification and valuation of network effects based on a deeper understanding of network characteristics and dynamics. This goal is achieved through a customer network-centric perspective to valuation in software markets. It allowed to develop a complex networks valuation framework for software markets based on the developed complex networks adoption and diffusion simulator. Moreover, the investigations provided a variety of further results:

1. *Identification of Network Effects in Software Markets.* The first research objective was to identify the role of network effects in software markets. Our investigations confirmed that network effects are vital determinants of software market dynamics which are currently not adequately represented.[3]
2. *Valuation of Network Effects in Softwares Markets.* After the identification of network effects, another objective of this research was to integrate them into valuations in software markets. Therefore, we determined the optimal valuation approach that accounts for network effects and integrated the insights on network effects with those on valuation.[4]
3. *Quantification of Network Effects in Softwares Markets.* Our research revealed that the quantification of network effects is the key to the integration of network effects into valuation in software markets. Hence, we designed software market models in order to derive additional quantitative insights on network effects in software markets.[5]
4. *Valuation Framework for Valuations in Software Markets.* In the next step, we developed a network effects framework for customer network-centric valuation in software markets that accounts for network effects.[6]
5. *Network Topologies and Network Dynamics in Software Markets.* A reconsideration of the developed network effects framework emphasized the vital importants of network properties and dynamics. While conventional research assumes frequently homogeneous random customer networks, complex networks research

[2] Please confer Sect. 1.1.

[3] Please confer Chap. 3 and Chap. 5.

[4] Please confer 4.

[5] Please confer Chap. 7 and Chap. 11.

[6] Please confer 4.

provides an array of additional concepts and insights on network properties, topologies and dynamics. Hence, we assumed a complex networks approach to valuation in software markets.[7]

6. *Bridging the Research Gap between Social Sciences and Natural Sciences.* Despite of initial progress, there remain considerable gaps. Based on the chosen research approach, we revealed the opportunities and limitations of an interdisciplinary complex networks research approach. Hopefully, this contributes to bridge the existing research gap between social sciences and natural sciences.

All in all, the results indicate that the investigations of network effects and the complex networks perspective can contribute to an increased quality of valuations in software markets as customer networks provide additional network theoretical insights. These come at the cost of additional efforts for collecting the required data, but, the insights can be worth the additional efforts, particularly in turnaround financing valuations. As, however, the required data is not always available, it is not always reasonable to conduct a network analysis. Hence, it is necessary to decide on a case by case basis if it is reasonable to pay for the additional insights. Although the outlined approaches are not tools to justify each and every investment, they enable contracting parties to negotiate investments in software markets based on capital market rationales. A comprehensive network analysis provides additional information in the assessment of the risk-return trade-off and supports decision making in corporate turnarounds. The investigated approach reveals that a customer network-centric valuation approach allows one to explain the deviations between observed values and fundamental valuation results, but still requires several refinements and a considerable amount of information for mass real world implementations. Thereby, the outlined irrational behavior in turnaround financing of software companies is overcome and all contracting parties are better off. But standard valuation tools will remain the primary choice of practice and further research is required. But in the case of high uncertainty and high flexibility it is reasonable to consider the benefits of a customer network-centric valuation based on the complex networks adoption and diffusion simulator.

[7] Please confer Chaps. 10 and 13.

Chapter 16
Implications of Results

"Software companies caught in a downward spiral find it exceptionally difficult to escape. Yet a few determined succeed."

(Blumling et al. 2002)

While the pursued customer network-centric valuation approach contains a variety of implications, this chapter summarizes the most relevant issues with respect to the outlined target group. First, findings relevant to managers of companies operating in software markets are presented, before implications for financial sponsors are depicted. Then, the consequences for various streams of research conclude this chapter.

16.1 Implications for Management

In this first section the most relevant implications for managers are investigated. All in all, managers should note that the outlined approach supports decision making in software markets based on a better understanding of customer networks, their topologies and their dynamics. This has implications on various aspects of management, such as strategic management, marketing and turnaround management. It is important to note that the following categories are not mutually exclusive and collectively exhaustive as some of the following implications are relevant to more than one cluster.

16.1.1 Implications for Strategic Management

The findings have a variety of implications on strategic management as it generated new insights by combining theoretical models, empirical data and numerical simulations. In the following, the most relevant implications are depicted concerning the modeling of product diffusions in software markets, the software design strategies, mega institution management and innovation portfolio management.

A. Kemper, *Valuation of Network Effects in Software Markets*,
Contributions to Management Science, DOI 10.1007/978-3-7908-2367-7_16,

Modeling Product Diffusions in Software Markets

The analysis reveals that the customer network is one of their most valuable assets of software companies. With the outlined method is can be approximately valued. For this reason managers should note that the outlined approach supports decision making in software markets and that any management strategy should aim at supporting this very important asset. Moreover, it is important to note that the modeling task should be performed by experienced professionals familiar with such complex networks and the respective valuation literature on the other hand. New insights from the theory on how networks develop can be gained by applying the developed model to strategic management decisions and marketing strategies. However, management will not be likely to obtain the information required and knowledge by themselves.

Software Design Strategies

The investigations reveal that compatibility is a key lever of strategic management to position products in software markets due to network effects. This finding is confirmed by research with similar findings (Wiese 1990). If the design of two products is incompatible, competitors are confronted with the start-up problem and with inter-sectoral competition. In the case that both rivaling suppliers design compatible products, they reinforce each other to overcome the start-up problem in an innovative market, but at the same time they are also in intra-sectional competition. The research illustrates the underlying fundamental trade-off between openness and control. Open standards increase the probability that a certain standard dominates a markets, but at the same time limits the opportunity to control the market development. In turn, a high degree of control is likely to delay the product diffusion (Shapiro and Varian 1998). Therefore, software companies have the choice to face more intense competition in larger markets and less intense competition in smaller markets. While compatibility relaxes the competitive pressure in the early stages of the product-life cycle, it intensifies the pressure in later stages (Katz and Shapiro 1986). At the same time, compatibility changes the nature of competition as consumers gain the flexibility to combine products from various suppliers. In turn, some companies introduce incompatible products in order to create a temporary competitive advantage that is used in order to set the market standard. As consumers try to avoid stranding costs, it can be rational for competitors to agree on a common standard in order to jointly develop the market based on a common installed base.[1]

Mega Institution Management

Empirical studies investigating the distribution of company turnover in various industries observe a trend towards mega institutions (Zanini 2005). In other words, the distribution of companies according to their turnover follows a powerlaw.

[1] A recent example is the decision of Toshiba to abandon the development of the HD-DVD standard in order to boost the development of the High Definition format.

Moreover, a comparative-static analysis, comparing the distribution of companies from 1994, 2007 and 2008, reveals that this mega institution trend is reinforced over time. In this study, the distribution of companies with respect to their size is compared for various dates. The comparison identifies increasingly steeper power curves. This means that few companies are extremely large, while many others are very small. While this phenomenon can be identified in various sectors, it is primarily evident in intangible asset intensive industries, such as software markets. In addition, it is interesting to note that the domination of few dominant companies is reinforced by the economic crisis of 2008 (Zanini 2008). The studies reveal that such developments are particularly relevant for companies operating in software markets (Zanini 2008). This phenomenon can be interpreted and explained based on the conducted investigations of this research. Our research revealed the outstanding role of customer networks in software markets, which are governed by network effects. From a customer network-centric perspective, software companies generate disproportionately high revenues if they benefit from strong customer networks that provides them with a dominant market position. If the network effects are strong enough, some already large companies are transformed into mega-institutions that dominate the business landscape, e.g., Microsoft (Zanini 2005).[2] Hence, the developed complex networks framework can be applied in order to generate a numerical network model that can be used to simulate the dynamics of the complex customer network. Such investigations allow a better understanding of the industry dynamics which, in turn, is a preqrequisite for determining the optimal strategy in the management of mega institutions.

Innovation Portfolio Management

Manager can apply the complex networks framework in order to optimize innovation management. A fundamental problem of innovation management is to rationalize investments in innovations. Although innovations are generally perceived to be positive, value tracking, controlling and valuation of innovations is challenging. In this context, the complex networks framework for valuations in software markets closes a research gap. Based on the customer network-centric perspective, innovation projects can be interpreted and valued as innovation options. From this real options perspective, a company is a portfolio of individual projects that are options on cash flows. Since the various options interact with each other, portfolio management strategies can be applied in order to optimize the performance of the innovation portfolio. A prerequisite is that it is possible to quantify the value of an innovation option. Based on the complex networks valuation approach, innovation options can be valued in order to determine the optimal innovation strategy based on transferring insights from portfolio management, e.g., risk diversification. Moreover, if innovations are interpreted as innovation options, it is possible to determine a return on

[2] Please confer Sect. 5.3.4.

the innovation by comparing the investment costs with the respective returns of the innovation. From this perspective, the venture capital market can be interpreted as a market exchange for innovations and financial investors can be seen as innovation brokers. They can apply the developed complex networks framework in order to optimize their innovation portfolio management.

16.1.2 Implications For Marketing Strategies

The research has also implications on marketing strategies of software companies. The systematic complex networks approach can be applied in order to develop an integrated marketing and financing strategy that can be communicated to customers and financial investors. The following approaches allow one to apply the insights on network characteristics and dynamics into normative marketing strategies in the following subsections.

Guerilla Marketing Strategy

Guerilla marketing is an unconventional marketing measure that creates social interactions with respect to a product, a brand or a company based on the surprise effect (Levinson 1984).[3] Its forms are manifold, but share the principles of guerilla tactics such as rebellion, surprise, and efficiency. Such actions are not expensive, and do not use classic channels of communication. Norms are broken in order to create a surprising effect for crossing the recognition barrier of the target audience. Successful guerilla measures are innovative and unique. Depending on varying degrees of media resonance, three subforms of guerilla marketing are distinguished.

1. *PR-Guerilla.* The intention of PR-guerilla is to get attention through controversial actions (Zerr 2004). The media resonance resulting from such actions is relative to other news and bears the risk that journalists respond with a negative feedback (Koppelmann 2006).
2. *Live-Guerilla.* Live-guerilla actions are based on a live performance that intends to stimulate the attention of potential customers. A product, brand or company can be the focus of the action or just a part of the performance. Examples are professional actors playing satisfied customers or hired chatroom-users.
3. *Street-Guerilla.* The purpose of Street-guerilla is to use unconventional methods to gain the direct attention of potential customers. An example is a beer brewery that delivers the beer through the front door during main customer hours in order to incite discussions.

From a complex networks perspective, all three Guerilla marketing strategies can be reasonable options in software markets. As a vital goal of Guerilla marketing is

[3] Guerilla markting is also coined Street-, Grassroots-, Ambush-, or Buzz-Marketing.

to foster, with minimal resources the social interactions in the customer network, it is necessary to define the exact communication goals and to test the effects of the planned actions in the network. Guerilla tactics can only be successful in software markets if the recognition boundaries of potential software users are crossed. Therefore, a campaign should be designed to primarily target opinion leaders of a customer network in order to maximize the impact of the campaign.

Viral Marketing Strategy

Viral marketing describes the development of entertaining or informative messages that incite receivers to pass it via primarily digital distribution channels to other potential customers or existing consumers (Rayport 1996; Groeger 2008).[4] The goal is to benefit from the network effects of digital word-of-mouth in social networks (Watts and Preretti 2007). Successful campaigns reach several million users within a short-time horizon, and are based on low spreading loss through the filtering in the social network among the nodes of the networks.[5] A prerequisite is a target group oriented design of the campaign that balances creativity, provocation and the product or company message. In addition to personal communication videos, audiofiles, animations, documents, online games and microsites are also possible distribution channels.[6]

From a network perspective, the following aspects are particularly relevant for a viral marketing strategy in software markets. The goal is to foster the social interaction among the customer network based on multi-stage referrals. Therefore, it is necessary to formulate a clear message that shall be transported. In software markets, the focus of such a campaign will be on the software product or the respective software company, and there frequently will be no direct interactions. It is also possible to combine this strategy with other measures, such as a combination of a guerilla strategy that is propagated based on viral marketing. In software markets, the viral marketing message can contain entertaining or informative content. The success of such measures is determined by the personal relevance, the emotional surprise and it can maximize its impact if the respective opinion leaders are targeted. An integrated approach requires that supportive additional measures are executed. But although an increasing budget is likely to provide an increasing level of flexibility, it is not possible to guarantee the success of a viral marketing strategy, particularly in software markets, as individual factors such as fads, trends and moods influence the outcome of a viral marketing campaign.

---

[4] The research on viral marketing dates back to a publication of Rayport in 1996 (Rayport 1996). While there are no coherent definitions of the term, the concept is also known as Virus-Marketing.

[5] The Quiksilver Dynamite Surfing campaign is an example that in 2007 reached a million users during the first week after its launch, and more than 10 millions users within the first 2 months.

[6] Microsites are personalized websites with a focus on a specific topic for a target audience.

Seed Marketing Strategy

A seed marketing strategy suggests that a company provides limited or unlimited access to one or multiple products for a selected group of test users for free or at a discount in order to foster social interactions between them and their environment with respect to the product (Solomon and Rabolt 2004; Groeger 2008). While the give-away of free samples is not an innovative concept, seed-marketing strategies do not aim at a general public audience, but primarily at selected opinion leaders.[7] The goal of such a strategy is to overcome potential financial barriers of the opinion leaders, to allow them to identify with the product and to benefit from the endowment effect, according to which people value a good or service more once a property right to it has been established (Kahneman et al. 1990).[8] The concrete implementation of the seed marketing strategy depends on the characteristics of the product, the goal of the campaign and the company. It is frequently related to additional background information on the product. Thereby, the opinion leader have more information than their social environment and can use it in order to underline their opinion leader status. The feedback of the opinion leaders provide additional possibilities for social interactions. As it is possible to combine various network marketing strategies, it is also possible to invite the participants of a Guerilla event to participate in a seed marketing campaign by inviting them with free samples of the promoted product.

From a complex networks perspective, a seed marketing strategy intends to foster the social interactions of the nodes in the customer network based on personal referrals by opinion leader. It can comprise personal as well as digital interactions through direct personal experiences, and thereby allows a strong identification of the opinion leader with the software which is not typical of classical marketing strategies. It seems like an old concept with a new label, but the organized design of a multiple stage referral strategy based on personal experiences allows a new dimension of involvement and commitment of the participant which increases the probabilities of a reinforcing feedback loop in the customer network. If the seed marketing strategy successfully creates the experience of personalized exclusivity and an advanced level of information, the participants are more likely to propagate the software in their social environment and, thereby, emphasizes their role as opinion leader.

User Generated Marketing Strategy

The idea of user generated marketing is to actively involve the consumers in the design of the product development and marketing strategy (Groeger 2008). An extreme example is the product development process of Brewtopia, an Australian

---

[7] Please confer (Solomon and Rabolt 2004) for further details on the distinction between product placement, free samples and seed marketing.

[8] A popular example is the invitations of Hasbro for 1,600 kids between eight and thirteen, who are considered to be cool alpha pups in order to let them test unreleased videogames with the twofold purpose of gaining an upfront feedback of highly relevant customers and to use them as multiplicators through the artificially created exclusivity.

brewery, which created a previously non existing product in a contested market without any investments into marketing, without its own production, without personnel and without capital. The founder invited 140 people to a 13-week workshop with the goal of designing a new beer, to find a name for the beer, to determine a fair price and to discover optimal distribution channels in exchange for one share per vote. Based on this user generated beer design the company reached 16,000 participants, sells beer in 50 countries and went public in 2006, 4 years after its foundation (Ramge 2008). The idea behind this strategy is to create a strong relationship with the consumer based on recognition and entertainment which is shared with friends and families. This creates a strong product related message that is spread in the social environment and incites others to join the network.[9] Such an decisive influence from the customer on entrepreneurial decisions is seen from a variety of perspectives. While some researchers emphasize the loss of control in the design of the marketing strategy which can lead to a bluring or even opportunistic image of the company, other research identifies the user generated marketing as a necessary step resulting from a paradigm shift in marketing.[10]

A user generated marketing strategy in software markets intends to incite opinion leaders of the software to participate in the design and marketing of the software. But while it is very risky to transfer the full responsibility to the customer, it is also possible to consider an increasing degree of participation. Accordingly, existing social interactions with the customer network should be monitored closely and extended if this is reasonable. However, it is important to note that this strategy is not suitable for all software products and companies. The increasing degree of participation can be distinguished in the following steps. In the first step, it is possible to integrate customer contribution into the marketing strategy, e.g., by customer generated commercials, Internet sites or jingles. In another step, a company may integrate suggestions of customers with respect to the design of the product packaging, which, however, is not a crucial aspect in software markets. Moreover, it is possible to integrate customer contributions into the product development of existing as well as into the product design of new software. This escalating involvement approach allows a company to integrate the customers actively into the user generated marketing strategy, while the control over the software remains with the software company. A comparison with the seed marketing strategy reveals that the user generated approach should be preferred if the customer network is larger, and if the relevance of the individual hubs is lower. It provides, however, the opportunity to discover opinion leaders and to integrate them into the customer network through their higher involvement if they have the corresponding incentives.

Vaporware in Software Markets

Expectations are vital determinants of diffusion processes in network effect markets that have to be carefully managed (Choi et al. 2005). As previously outlined

[9] This phenomenon is also known as the Hawthorne-effect. Please confer (Jones 1992) for further details.

[10] Please confer (Groeger 2008) for further details on this discussion.

autoregressive market expectations of customers shape the competitive landscape and can be influenced by investments in research and development, human capital, and infrastructure, as they are interpreted as credible commitments of companies (Besen and Farrell 1994).[11] This relationship emphasizes the importance of information management and investor relations. Announcements of product launches, or of compatibility influence the purchasing decision of customers and, thereby, the future development of the company (Swann and Shurmer 1994). In this context it is interesting to note that it is common practice in software markets to pre-emptively announce vaporware in order to increase the entry barriers that deter potential rivals to enter the market.[12] A review of the empirical literature reveals that it is a widely adopted practice to announce innovations well in advance of actual market availability. Accordingly, firms announce innovations in order to convince market participants that their own innovation will become a market standard and, thereby, to reduce the perceived uncertainty of potential customers. Incentives for such pre-announcements are even stronger in markets with network effects, since the announcements are used to reduce the delay of purchases and to sabotage the build-up of competing installed bases. If such announcements are not fulfilled, however, the credibility of the company is damaged which is harmful for long-term development. Therefore, management of the customer's expectations is a prerequisite for a long-term success in software markets.

Opinion Leadership in Software Markets

The opinion leadership concept assumes heterogeneity of agents in a network, and states that certain participants of the network have a higher influence on the adoption decision of others than the average agent (Valente and Davis 1999). Hence, they should be the primary target group of marketing activities as they help to overcome the market introduction phase more quickly, allow for a longer and broader harvesting in the growth and saturation phase, and establish customer loyalty in the decline phase. This insight leads to the design of optimal market penetration strategies, such as the two-step flow strategy according to which opinion leaders are targeted first, in order to convince, and in the second step the market followers as selective individuals with a high centrality have a high influence on the decision of other participants. A successful example of such an opinion leader marketing is the pharmaceutical industry which spends approximately 32% of the marketing budget for attempting to influence the opinion leaders of social networks. Similarly, the analysis of customer networks in software markets revealed that they frequently have a scale-free topology. There are some nodes which have a higher degree than other actors. Consequently, an optimal allocation of resources requires one to segment the potential customers into various strata according to their connectivity potential, and to focus marketing efforts on these opinion leaders according to a hierarchy of customers (Shapiro and Varian 1998). In scale-free networks there are two classic levers for

[11] Please confer Sect. 5.3.1.

[12] Vaporware are products which are not yet ready for a market launch.

growth. The first one is to back a viral adoption of the network by relying on the scale-free network topology, i.e., every new user of a network is potentially a new node of the social network structure. Hence, it is necessary to develop and utilize respective tools that induce them to bring all their real-world connections into the virtual community. Second, it is important to support the animation of the network as a central hub may not be an animator for other customers. In this case, animation is essential in order to convince other customers to buy the software. But in order to pursue this strategy, it is necessary to identify opinion leaders in real-world networks. Innovative results can be achieved by combining ideas of epidemiology with those of network resilience, similar to vaccination strategies. From a complex networks perspective, vaccination can be interpreted as the removal of some particular set of vertices from a network, which in turn can be modeled as site and bond percolations on networks (Callaway et al. 2001). If the site percolation is correlated with the vertex degrees of networks, the structure and function of complex networks it is possible to develop targeted vaccination strategies, e.g., by removing vertices with the highest degree (Pastor-Satorras and Vespignani 2002). As networks tend to be particularly vulnerable to the removal of their highest degree vertices, this kind of targeted vaccination is expected to be particularly effective. However, it is not easy to identify the highest degree vertices in real-world social networks. With respect to software markets, this goal is equivalent to the problem of identifying opinion leaders in customer networks. An interesting approach to solving this problem is based on the observations that since the probability of reaching a particular vertex by following a randomly chosen edge in a graph is proportional to the degree of the vertex, it is more likely to find high-degree vertices by pursuing edges than by choosing vertices at random (Cohen et al. 2002). Consequently, research suggests that a population should be immunized by choosing a random person from that population and vaccinating a friend of that person by asking about a personal relationships with a higher degree and then repeating this process (Newman 2003b). Analytic calculations and computer simulations reveal that this strategy is substantially more effective than random vaccination (Cohen et al. 2002). Contact tracing methods used to control sexually transmitted diseases and ring vaccination method used to control smallpox are both examples of such vaccination strategies (Kretzschmar et al. 1996; Mueller et al. 2000). Such insights on vaccination strategies can be transfered, within the outlined limits, to software markets. Accordingly, a random customer is chosen from the customer network and asked to recommand a neighbor with a higher node degree. If the customers cooperate, this search strategy is far more efficient than analyzing the total customer network. A crucial issue of this strategy, however, is to provide the incentive for customers to participate in the referral process which is an open research issue for marketing research.

Pricing Strategies in Software Markets

Network effects are pervasive in software markets. While software vendors consider pricing strategies, they also must take into account the impact of network effects on sales as they can be used to protect or to extend the market share, e.g., due to the start-up problem (Farrell and Saloner 1986).[13] Pricing strategies are primarily applied to satisfy one of five common strategic objectives (Besen and Farrell 1994; Wied-Nebbeling 2003).

1. Profit maximization
2. Achieve a target return on investment
3. Achieve a target sales level
4. Achieve a target market share
5. Prevent or influence competition

Accordingly, factors that determine pricing policies can be summarized in the following four clusters (Varian 1993; Besen and Farrell 1994; Shy 2001):

1. *Customers.* Consideration of customer expectations about price must be addressed. Empirical research reveals that there exist psychological pricing barriers that have to be taken into consideration. Ideally, a business should attempt to quantify its demand curve to estimate what volume of sales will be achieved at given prices.
2. *Competitors.* If the business is a monopolist, then it can set any price. At the other extreme, if a firm operates under conditions of perfect competition, it has no choice and must accept the market price. In most cases the market entry and exit barriers are in the middle. Hence, the chosen price needs to be very carefully considered relative to those of close competitors.
3. *Costs.* While in the short-term it may be acceptable for a company to price a product below total cost if this price exceeds the marginal cost of production, a business model has to ensure in the long run that its products are priced above their total average cost. In addition, it is also important to consider learning curve effects and economies of scale. These vary along the product life cycle and are vital determinants of the optimal pricing strategy, due to multiplicator and obsolence effects.
4. *Business objectives.* A fourth factor is the outlined business objectives. These are comprised of maximizing profits, to meet a specific target performance level, to achieve a target sales level, to influence the market share or the competition.

In order to achieve these objectives based on the respective factors, a variety of pricing strategies are available along the product life cycle for competition in software markets (Pechtl 2005):

1. *Marginal cost pricing strategy.* According to the marginal cost pricing strategy, prices are set in relation to the variable costs of production. Its objective is to achieve a contribution towards fixed costs and profit. Hence, prices are set using variable costing by determining a target contribution per unit which reflects a

[13] Pricing strategies are defined as constitutional decisions with respect to pricing parameters under consideration of long-term effects (Vahlen 2006).

target variable cost per unit, total fixed costs or a desired level of target profit. The marginal cost pricing strategy is suitable for short-term decision-making, avoids any arbitrary allocation of fixed costs as well as overhead. It narrows the focus of the business towards a break-even perspective that is pretty common in software markets. Nevertheless, there are some potential disadvantages of using this method, such as the risk that the price set will not recover total fixed costs in the long-term, and it may be difficult to raise prices if the contribution per unit is set too low.

2. *Cost plus pricing strategy.* The full cost plus pricing strategy aims to set a price that accounts for all relevant costs of production. In comparison to the marginal cost pricing strategy, it has a variety of advantages as price increases can be justified due to rising costs, and a price stability may arise if competitors take the same approach and have a similar cost structure. In turn, some disadvantages are that the cost plus method ignores the price elasticity of demand, i.e., it may be possible to charge a higher price to maximize profits, the business has less incentive to save costs, it requires an estimate of business overheads and it may leave a business in a vicious circle.
3. *Penetration pricing strategy.* Penetration pricing is based on setting lower, rather than higher prices in order to gain a large, if not dominant market share. Market penetration strategies differentiate between opinion leaders and market followers and are primarily applied if a new market is entered. The expectation is that a quick product diffusion allows to benefit from cost degression that allows a market leadership position. Popular examples of penetration pricing are Internet browser, such as Netscape, business networks, such as Xing, Facebook, or m-Commerce applications such as Paybox. All products have in common that the products are introduced for free, before the price, at least for specific services, are increased. Penetration pricing is frequently considered to be an optimal for software markets in order to gain a critical mass of customers, while other empirical examples contradict this claim, e.g., the failure of Borland's discount strategy (Farrell and Saloner 1986). Hence, it is only possible if the demand for the software is highly elastic, i.e., demand is sufficiently price-sensitive. If applied successfully, penetration pricing may lead to large sales volume and relatively low costs per unit. Economies of both scale and experience allow lower production costs, which justifies the use of penetration pricing strategies to gain market share. But a penetration pricing strategy is also frequently applied in order to promote complimentary products. While the main software may be priced with a low mark-up to attract sales, customers are then required to buy additional products which are sold at higher mark-ups. A crucial issue is that a supplier must be certain that it has the production and distribution capabilities to meet the anticipated increase in demand before a penetration pricing strategy is implemented. But due to the large scalability of software, such penetration strategies are frequently applied in software markets. A potential disadvantage of penetration pricing strategies is the possibility that competitors copy the strategy by reducing their prices too, thus nullifying any advantage of the reduced price. In

addition, there is an impact by the reduced price on the image of the offering, particularly where buyers associate price with quality.
4. *Price skimming strategy.* Price skimming strategies imply charging a relatively high price, at least temporarily above the long-term optimal price, for a short time if a product is launched into a market, before it is decreased subsequently. A classical example is the pricing strategy of Intel for new processors. The objective is to differentiate the willingness-to-pay of customers who are willing to pay more for having the product sooner, as later prices are lowered when demand decreases. Its success is largely dependent on the inelasticity of demand for the product, either by the market as a whole, or by certain market segments. Thereby, the supplier benefits in the short term from monopoly profits. But depending on the network effects which are barriers to entry additional competitors are likely to be attracted to the market with increasing profitability. Hence, prices will fall as soon as competition increases. Thus, the main objective of a price-skimming strategy is to benefit from high short-term profits due to market segmentation. Price skimming has a variety of advantages such as high returns on the high set-up costs, a high-quality image, effective segmentation of the market and higher mark-ups for dealers.
5. *Expansionistic pricing.* An expansionistic pricing strategy is a more drastic form of penetration pricing as it implies setting very low prices in order to establish mass markets. It is suitable if the market is characterized by a high price elasticity of demand, such that the adoption of a very low price implies a significant increase in sales volumes. The strategy is applied for reasons similar to the penetration pricing strategy. Expansionistic pricing strategies may be applied by companies attempting to enter new or highly contested markets. Additional lower-cost versions of a product may be offered at a very low price to gain recognition and acceptance of consumers. As soon as acceptance has been achieved, more expensive versions or models could be offered. Markets that might benefit from expansionistic pricing strategies include magazine and newspaper publishers.

In general, the relevance of pricing strategies are market- and company-specific, e.g., relative size, competition, cost structure, etc. Nevertheless, it is possible to derive some application guidelines for software markets from a complex networks perspective. Cost-based pricing strategies are not an option as the costs of reproduction are close to zero. Instead, software companies frequently choose a customer-value strategy which is based on the assumption that the product provides a specific value to the customer that justifies a differential pricing strategy (Shapiro and Varian 1998). Alternatively, skimming, penetration pricing and entry limit pricing are possible (Späth 1994). In addition to the outlined pricing strategies, a technology can be sponsored or unsponsored (Besen and Farrell 1994). If a sponsored strategy is pursued, the software company subsidizes the diffusion of a technology. Such a sponsoring may be profitable if it is reasonable to increase its popularity, to increase its market share or to cross the critical mass in a market. Unsponsored technologies, in turn, are sold based exclusively on the original and derivative utility generated by product. Further strategic options are to reduce the marginal costs by process innovations, or to raise the market entry barriers for competitors by monopolizing restrictive production factors, e.g., by increasing investments in research and development. This has an impact on the dynamics of the customer network, if the respective process innovations allow a decreasing indirect costs which in turn increases the probability that the software crosses the critical mass of customers. In summary, the software firm will charge a lower price to attract more users in the first stage, before it increases the prices of the software. If multiple product generations are compared, a dynamic analysis reveals that the optimal upgrade time is when gross profit of the first software edition equals the gross profit of the second edition (Yang 1996). Hence, too early or late promotion of the new edition will cause profit losses. The findings imply that the complex networks adoption and diffusion simulator can also be applied in order to support the determination of pricing strategies and the optimal upgrade time.

16.1.3 Implications For Turnaround Strategies

The findings of the research reveal that investigations of customer networks contain a variety of information that can be used for management decisions. They have in particular implications for turnaround assessments in software markets. Corporate turnaround decisions are frequently irreversible investments, but the research indicates that it can be rational to invest into a software company despite of negative operative cash flows. It can be a value-maximizing strategy to interpret temporary losses as an option premium in exchange for potential future pay-offs from a real growth option. In this context, the suggested complex networks valuation framework enhances the transparency and the quality of valuations based on which the turnaround decisions are made. Therefore, accurate valuations in software markets are a central element of efficient turnaround assessments, an aspect which is not covered sufficiently by existing financial research. Consequently, irrational investment decisions occur due to the underinvestment problem as the intangible but valuable customer equity of customer networks is frequently ignored. But, since the customer base is identified as a central strategic variable in competition, it may be reasonable to extend the installed base despite of temporarily negative cash flows, if the overall value including the implied managerial flexibilities is positive.

Intertemporal Resource Allocation

Companies in turnaround are frequently in a dilemma. On the one hand operative short-term measures are required in order to assure the operative business, while on the other hand it is important to finance medium- to long-term projects in order to assure the sustainable success of the company. Research on turnaround reveals that successful turnaround strategies require successful management of both. It is important to provide the short-term liquidity in order to keep up the operational business, but it is also important to allocate enough resources into the research pipeline.

In this dilemma a crucial aspect is to convince internal and external financiers that investments, e.g., into the customer network are not voluntary, but are as necessary as operative funds, e.g., if they are required in order to reach a critical mass of customers within a short-term time horizon. In such negotiations the complex networks framework for valuation in software markets can bridge the gap between both parties by providing a theory-based communication tool that allows one to discuss the underlying assumptions, but provides a platform for illustrating the intangible value of such software companies as the concepts also have ontological implications. It provides both parties, the financial investor and the management, with a common framework which is the basis for the negotiations. The model is not supposed to rationalize each and every investment, but to provide a common language for the discussion of the significant underlying problems. The perception of individual parameters and their values will be different, but the parties will be enabled to exchange information within the synchronized mental model and, thereby, to focus the negotiation process on the controversial underlying assumptions.

Undervaluation Problem

The research reveals that traditional implicit approximations of customer networks in corporate turnaround assessments are frequently inadequate. They tend to underestimate the value of the analyzed company as it is necessary to consider intangible customer networks as intangible but valuable assets of software companies. This implies that a reliable turnaround assessment has to account for network effects. Consequently, it is rational to invest into the customer basis in the short-run despite of negative operative cash flows as such investments can be decisive to reach a critical mass of customers. Undervaluation, in turn, implies that some companies are not considered for turnarounds. In this case, the ignorance of network effects and complex networks leads to inefficient turnaround decisions based on biased assessments. Credible turnaround assessment models account for network effects of customer networks, as they are supposed to increase financial transparency and to avoid inefficient capital budgeting decisions. While traditional approaches have several limitations, alternative real options approaches provide a platform to frame strategic and operative flexibilities. Hence, it can be a value-maximizing strategy to accept short-term losses in software markets and to interpret them as an option premium for potential future pay-offs.

Cost Cutting Potential

A complex networks analysis can illustrate a variety of cost cutting potential. Traditionally cost reductions are achieved by saving personal expenses. But another important aspect is the role of customer retention that contains a variety of cost cutting potential. First, the efficiency of retained customers is greater as retention costs are frequently lower than the costs of acquiring new customers (Rust et al. 1995; Blattberg and Deighton 1996; Sheth and Parvatiyar 1995). This implies that even higher marketing investments are necessary if the company performance is low and requires a turnaround. These increased investments have to be financed with financial resources. Such insights are particularly important in the design process of a turnaround strategy, as marketing costs will all else being equal decrease with an increasing success of the company. Moreover, it is necessary to consider a focus on the most profitable clients. From a complex networks perspective, such customers are the central hubs of the customer network who should be targeted in order to optimize the allocation of resources.

Integrity of Internal and External Corporate Networks

A vital result of respective network theoretical investigations is that the integrity of internal and external corporate networks is extremely fragile in turnarounds. Investigations on the resilience of networks reveal the vital importance of network integrity (Hommel and Kemper 2006). Similar to positive reinforcing effects, even a few central nodes can threaten the integrity of internal and external corporate networks with rebalancing negative effects, e.g., by eroding a loyal customer basis. Due to hypercompetition, such attacks can have significant implications on the performance of the company. Depending on the circumstances, it is possible that internal and external networks are disintegrated inciting, which in a worst case scenario can cause the liquidation of companies (Hommel and Kemper 2006). Percolation models can be applied in order to test the resilience of customer networks in software markets, as they allow to calculate implications of a random edge deletion. Such features are particularly useful for observing the effects of targeted attacks on the average path length of networks as in the cascading failure model (Holme et al. 2002). Accordingly, if networks are restricted by a maximum carrying capacity per node, the failure of a node implies a redistribution of its load on neighbor nodes. This can cause a cascade of subsequent collapses, if the network operates close to its carrying capacity. Hence, the failure of a single node can result in the total collapse of a network. Applied to software markets, the complex networks software market models developed with the simulator can be applied in order to investigate the resilience of customer networks. For this purpose, various destabilization strategies could be applied, e.g., to delete randomly customers or central customer, in order to assess the stability of the customer network. Such network theoretical information on the customer networks are highly relevant in order to develop sustainable business models.

16.2 Implications for Financial Sponsors

In this section, the focus is on implications for the business of financial investors, investment banks and venture capitalists. Although some of the concepts are increasingly recognized in the corporate finance practice, the overall popularity and impact is rather limited, despite a broad variety of implications (Mauboussin et al. 2000). Hence, the most relevant aspects for financial sponsors are highlighted in the following sections.

16.2.1 Business Plan Analysis

The analysis of business plans is at the core of many investment and financing decisions of banks and financial sponsors. In this context, the outlined complex networks adoption and diffusion simulator can provide a valuable contribution by testing the plausibility of business plan assumptions in order to rationalize the respective investment decisions. Such investigations are particularly reasonable as part of a commercial due diligence, i.e., an in depth market analysis.[14] If the commercial due diligence has a high deal relevance, e.g., as in a turnaround assessment, complex networks investigations based on the outlined adoption and diffusion simulator may contribute to a better understanding of the market. Consequently, it is possible to design more reliable market models as input factors for valuations and subsequent investment decisions.

16.2.2 Critical Mass Turnaround Financing Opportunities

Based on the outlined business plan application, another central implication for financial sponsors concerns the turnaround financing analysis of turnaround candidates. A large proportion of companies in the dynamic software markets struggled in the beginning of the century, and many vanished from the market. The reason being is that despite a desperate need for financial resources, a significant number of software companies are not capable of convincing frustrated financial investors of further investments. While it is difficult to achieve a successful turnaround in any industry, this is a particular challenge in software markets.[15] On the other hand, investments into turnarounds can be very profitable if they are picked wisely. Empirical research reveals that 59% of the observed turnarounds decrease in value, whereas some of the successful cases are listed among the 25 most valuable software firms of the world.[16] Among the remaining companies, some 15% were acquired by competitors or financial investors, and 13% stopped the decline in value, but were not able to significantly increase their financial performance (Beer and Nohria 2001). But the outlined effects can be captured only with highly complex models

[14] Please confer (Sebastian et al. 2005; Niederdrenk and Maack 2008) for further information on a market or commercial due diligence.

[15] Research on turnarounds indicates that more than 70% of turnarounds fail, whereas only 13% of software turnarounds are successful (Blumling et al. 2002).

[16] Examples are BEA Systems, Oracle, and Peoplesoft.

(Meise 1998; Hommel and Müller 1999; Kühn et al. 2000). Therefore, a financial analyst is in a dilemma between the accuracy of the valuation and the information costs, as a variety of expensive data is frequently required. As the quality of the input data determines the quality of the financial analysis, prohibitive elevated prices can prevent the application of the real options approach, despite additional accuracy. In turnaround valuations for turnaround assessments such additional information can be worth the additional costs, as frequently a unique decision has to be made based on which the future of the company is decided. Therefore, the incremental gain of information can be worth the additional effort. The discussion on the contribution of real options to the outlined research question reveals that the exploration of the underlying cash flows is a central open issue in the design of turnaround valuation. These depend on the respective market developments and require a comprehensive analysis of the underlying customer network. Essentially, financial sponsors are confronted in such situations with critical mass turnaround financing opportunities. Accordingly, the key question is whether it is profitable for the financial sponsor to finance the corporate turnaround, or whether it is better to reject it. This question can be stated differently from a network theoretical perspective: Is the probability that a critical mass of customers can be reached within a predetermined time horizon, with a specific amount of financial funds, large enough to compensate for the respective opportunity costs? Or, in other words, is the probability that the software company gains the required X customers to reach the critical mass in this market segment within the next 6 months, if the company is equipped with Y financial funds, large enough to compensate for the respective risks? In order to solve this key question it is necessary to investigate the critical mass of customers which, in turn, depends on the customer network topology as depicted in the previous analysis. Hence, the complex networks adoption and diffusion simulator could be applied in order to support the turnaround financing decisions of financial sponsors. The key to answering this question is to interpret the turnaround investment as a real option, and to apply the complex networks framework for valuation in software markets. This allows us to investigate the relationship between the total company value, including the real option component, due to network effects and the respective costs. If the total value is positive, the turnaround project should be pursued, while it should be rejected if the respective value is negative. This critical mass turnaround financing analysis can be applied to other industries as well, but due to its dynamic nature it is particularly suited to software markets. Therefore, the findings of the research are a particularly relevant to turnaround financing decisions in software markets.

16.2.3 Business Identification Tools

The previous research provides the possibility to develop business identification tools for investment banks as the volatility in certain network effect markets can be observed. Hence, potential targets can be identified. Characteristics of an interesting target are a strong position in customer networks of software markets, e.g., opinion leader, accelerating sales growth and low reproduction costs. Targets with these attributes should be investigated thoroughly in order to identify lucrative investment opportunities (Mauboussin et al. 2000). On the other hand, rapid cluster formation in customer networks limits the organic growth potential of expanding software companies, as switching behavior of customers in software markets becomes increasingly unlikely with increasing network effects. Such network theoretical constellations in software markets may help to identify targets for mergers and acquisitions. Then, the customer network assessments can be conducted for potential clients in order to identify suitable corporations that benefit from better positions in customer networks. Moreover, they have access to data that could be collected in order to derive other industry specific network characteristics, which could be used for future network assessments, or aggregated and sold to clients interested in such customer network investigations.

16.3 Implications for Research

The previous findings provide interdisciplinary implications, but also implications specific to the involved streams of research. In the following section, first some interdisciplinary implications are summarized, before these findings are reviewed with respect to the corresponding discipline.

16.3.1 Interdisciplinary Implications for Research

Research on quantum mechanics revealed that it is impossible to predict the future.[17] Nevertheless, it is important for managers and financiers to base their decisions on solid information and reasonable decision making tools. Probability theory and option pricing theory fill this void between the unrealistic quest for certainty and an odyssey in uncertainty. In this context it is important to note that the contributions of complex networks allow us to analyze uncertainty in network systems by providing some general insights into the mechanics of diffusions in networks. The probabilistic description of network diffusion, with the help of network theory, provides a means to condense the information about the uncertain future into a probability distribution, which is transformed with the help of option pricing theory. The complex networks approach allows us to derive additional network theoretical information at

[17] This insight is an implication of the Heisenberg uncertainty principle stating that values of special pairs of variables cannot both be known with arbitrary precision. In other words, the more precisely a property is known, the less precisely the other can be known, e.g., position and momentum. Thus, it is not possible to develop a deterministic prediction of the future. It is important to note that this principle is not a statement about the limitations of research, but rather about the nature of the system itself (Heisenberg 1967).

the expense of certain modeling costs which can enhance the information base for decisions if the required information is available. This trade-off between additional costs and additional information is not profitable in all situations, e.g., a small local bakery is not likely to set up a network diffusion model for the global bakery market. Nevertheless, this additional modeling and valuation effort can be worthwhile in specific managerial situations, such as turnaround financing decisions. Therefore, the outlined research is relevant on a conceptual level to multiple managerial decisions while it should be implemented only if the possible benefits can outweigh the related costs. Research, however, in such cases revealed that the outlined models can provide valuable additional information which in turn will hopefully contribute to better decisions. After this general review of interdisciplinary implications, the most relevant implications are classified in the following, according to their respective disciplines.

16.3.2 Implications for Financial Research

As there are a variety of implications for financial research, only the most relevant are depicted in the following sections.

Convergence of Finance and Marketing Research

Research reveals a variety of aspects that point towards a convergence of financial and marketing research. Solid marketing research is required in order to configure the complex networks adoption and diffusion simulator for a better understanding of the customer network. In essence, this customer network perspective is a vital bridge between marketing and finance. The predominant DCF valuation method discounts expected cash flows at a risk-adjusted and capital weighted discount factor in order to derive the present value of the valued object. Hence, the projected cash flows are the vital input factor of this model. The estimation of the cash flows itself, however, depends on sales projection which are traditionally based on implicit marketing assumptions. In the current business practice, sales projections are frequently approximated based on historic data. For this reason, it is reasonable to consider the outlined concepts in marketing research, in financial research, and at the respective boundary between both disciplines.

16.3.3 Implications for Network Economic Research

The research on complex network economics is a logical extension of the classical network economic literature. While network effects are increasingly understood, the complex networks perspective allows an application of statistical mechanic tools in order to extend the analysis, particularly of large-scale networks. This enables the modeling of larger networks and, thereby, increases the realism of the models. Depending on the development of the computing power, an application of the outlined concepts and programs on upcoming computers with even higher performance allows the development of even more realistic models and simulations. Moreover, the outlined research illustrates that the combination of research on network economics and on complex networks provides a variety of insights. Such research on complex networks economics has a focus on the implications of network effects on large-size complex networks. Based on the framework, further research on specific aspects of complex networks economics is required, as depicted in the next chapter.

16.3.4 Implications for Complex Networks Research

The outlined analysis is an application of complex networks theory to practical applications. Since practice and theory have always helped each other to develop, there are some important insights that should be considered:

Ubiquious Network Properties and Dynamics

A central finding of the outlined research is the ubiquious importance of network properties and dynamics in economic phenomena, as the respective concepts are highly relevant to a variety of problem classes. Complex networks research provides a systematic overview of relevant features, tools and their explanatory potential. In addition, the outlined investigations revealed some fundamental insights on the general nature of diffusion processes in social networks. The key to complex networks research is to identify the underlying relevant network backbone, i.e., the network structure that drives the dynamics relevant to the investigated research question, before the respective diffusion process is selected and calibrated. This can be as diverse as a total population, an aviation network, or observed occurrences of dollar notes (Hufnagel et al. 2004; Brockmann et al. 2006). But, as first impressions can be misleading, it is important to challenge the choice of the selected network and to reconsider other network types. An interesting example of the new class of probabilistic models is the complex networks study on the diffusion of SARS based on an analysis of the aviation network (Hufnagel et al. 2004). It combines a local stochastic infection model with a stochastic aviation network that exhibits, in the limit of large populations, deterministic properties. Accordingly, it is important to recognize the two stage stochastic-deterministic nature of diffusion processes in social networks. In the first stage, the diffusion frequently follows stochastic patterns as they occur at random places. But once the diffusion is in progress, at a certain stage the process is increasingly determined by deterministic rules. Together both effects exhibit a hybrid semi-deterministic and semi-stochastic process. The model delivers reliable results, and indicates that the forecasts of the geographical spread of an epidemic is possible if the relevant network is known and if its parameters can be reasonably approximated. Nevertheless, such experiments reveal that after the parametrization of the simulations, sensitivity analyses are required as depicted in the outlined complex networks framework.

Social Network Systems

The analysis emphasizes the importance of recognizing that social networks, such as customer networks, are different from scientific networks and, therefore, require network specific research that accounts for such particularities. Social networks are far more fragile and unpredictable due to psychological aspects. Although the general diffusion mechanics may be similar, or even the same, it is important to note that the processes in customer networks may diverge in different types of networks. The research provides an excellent example, based on which further research is required.

Scaling Properties of Networks

The implications of the findings are twofold. First, in general the dynamics of random networks can be simulated in adoption and diffusion simulations due to their invariance to scaling. Second, a suitable adoption and diffusion simulation of a small-world or scale-free network has to account for the scale of the investigated population, as they are not invariant to scaling.

Chapter 17
Research Limitations

> "Everything should be made as simple as possible, but not one bit simpler."
>
> A. Einstein (1879–1955)

The previous chapters revealed a broad variety of findings and implications. Nevertheless, it is important to note that the research is also constrained by a variety of limitations. As these constraints are very diverse only the most relevant limitations are selected in the following. The limitations are grouped into general limitations, limitations of financial research, and limitations of complex networks research. The constraints are the basis for further research opportunities that are derived in the following chapter.

17.1 General Limitations

Some of the outlined ideas are still under development. They will not replace, but rather complement, existing approaches to valuation in software markets. In this context it is also important to be aware of the following underlying general assumptions and limitations of the presented research approach in order to derive reliable results. A general limitation of the research results from the previously defined scale and scope of the analysis. The pursued company perspective is relevant for managers and investors, but from an economic perspective it would also be interesting to adopt a *wellfare-theoretical point of view*. For example, by developing incentive mechanisms in order to overcome inefficiencies of individual purchasing decisions due to asymmetric and imperfect information. This, however, is beyond the scope of this book. Moreover, it is important to note that is not the purpose of the approach to justify each and every investment. Instead, it is a communication tool that provides two contracting parties, e.g., software management and potential investors, with a consensus building model. If both parties agree on a model, they can discuss the respective parameters based on a common understanding. Since both parties have different incentive systems, they will have diverging opinions on the parameters of the model, but they share a common basis for negotiations. Hence, the presented approach is a *communication tool* to structure negotiations of transactions. But, the

A. Kemper, *Valuation of Network Effects in Software Markets*,
Contributions to Management Science, DOI 10.1007/978-3-7908-2367-7_17,

overall explanatory power of the model depends on the scale of analysis and the respective data. If only low quality data is available, the results have to be applied with caution, whereas high quality data is scarce and expensive. Besides these general limitations the most relevant constraints with respect to financial research and complex networks theory are summarized in following.

17.2 Limitations of Financial Research

In addition to the general remarks, the book reveals a variety of limitations with respect to financial research. A selection of the most relevant aspects is depicted in the following section. The model should apply real-world data for validation in a cased study. However, the preparation of the case study revealed a *data collection problem* for the relevant data on customer networks without the support of the respective company and customers. In general, a significant part of the information relevant to valuations can be extracted from annual reports, financial statements, or research reports. But the required information for a customer network-centric perspective on valuation is very confidential, e.g., relationships among customers. Hence, it is frequently difficult or even impossible to develop a reliable model of the relevant customer network. Moreover, it is important to note that the costs of the required data can be prohibitive. All investment valuation approaches are characterized by a *dilemma between accuracy and costs*. This is also relevant in the pursued research approach. It is not trivial to conduct all steps of the complex networks framework for valuation in software markets and to calibrate a complex networks adoption and diffusion model. As time is frequently a critical success factor of managerial decisions, experienced modelers are required. Hence, financial resources and time are required. The trade-off implies that it is necessary to assess case-by-case the benefits and costs of the complex networks approach. While it may be rational to conduct the complex networks approach in some cases, it can be better to approximate the impact of network effects in others, e.g., if the costs are prohibitive. In addition to the restrictions of financial research, there are also some limitations from a complex networks perspective that are depicted in the following section.

17.3 Limitations of Complex Networks Research

Despite the explanatory potential of complex networks research, there are several limitations and trade-offs which restrain such investigations. The most relevant limitations are depicted in the following. The analysis of social networks revealed their specific nature. A particularly important aspect is that a partial network analysis requires one to determine the *boundaries of social networks*, which because of their fuzzy nature can be difficult to delineate. The strategy to incorporate all agents of a network in a total system analysis may result in precise findings, but it is not applicable for all large-scale networks. In a partial network analysis, in turn, may result in biased findings. Therefore, it is necessary to either make a theoretically informed decision about the significant boundaries of the model based on statistical theory or to work with the total population. This emphasizes the importance of *data robustness* in a network analysis. Our findings on the scaling properties of networks suggest that the smaller the network, the less reliable are the simulations. A reason for this inverse relationship is that randomness can have a much larger effect on results in small networks. In order to account for this problem, it is necessary to conduct a sensitivity analysis that investigates a range of values and the sensitivities to the underlying assumptions. In addition, further advances in *modeling complex networks* are necessary, e.g., by investigating other customer adoption rules, in order to increase to scale of possible phenomena that can be explained. In this context it is important to note that the choice of the adoption rule depends on the specific context. While some adoption models are suitable for special forms of social interactions, e.g., the decentralized standardization rule for short-range interactions, the mean-field approximation is suitable to model long-range interdependencies. All in all, there is no one general solution for modeling social interactions in customer networks. Hence, it is necessary to analyze the context in order to determine the optimal network topology and the respective adoption rule. Finally, some of the research problems are subject to *computational boundaries*. Since the computational requirements of the complex networks simulator are high, there are computational constraints for investigations of large-scale networks. Hence, additional computational resources will be required in order to simulate network properties and dynamics of large-scale or even full-scale systems. However, it is important to note that the boundaries of technical restrictions decrease over time, whereas the computational performance of computers increases. In addition, alternative methods, such as cellular automata, allow to conduct complex networks investigations even in large-scale networks if the diffusion dynamics are very simple (Kemper 2006). All in all, there are a variety of restrictions. However, some of the limitations can be resolved by further research efforts which are depicted in the following research outlook.

Chapter 18
Research Outlook

"This is not the end. It is not even the beginning of the end. But it is, perhaps, the end of the beginning."

Churchill (1942)

The research indicates the potential of complex networks research to study a wide array of problems. For this reason, further research activities should be engaged despite of the previously depicted limitations. From a very general perspective, a lot of research effort is required in order to unify the existing fragmented parts into a coherent body of research. But in addition, further research is required in order to approach research opportunities in financial research and complex networks research. The most important research opportunities are depicted in the following chapter.

18.1 Financial Research Opportunities

The application of complex networks research to investment theoretical research problems in software markets reveals its interdisciplinars contributions to financial research. The following three research opportunities are most relevant to advance the existing boundaries of financial research.

18.1.1 Opportunities in Investment Valuation

As previous research illustrates the contribution of complex networks research to valuation in software markets, the insights should be applied in order to approach further open research problems in investment valuation. For this purpose, there are two possible research strategies. The scope of research can be generalized or specialized. According to a generalization research strategy, the complex networks approach is applied to other industries as well. In this context, it is important to note that the dynamic nature of software markets is particularly suited for the pursued investigations. However, it would be interesting, in turn, to study the implications

A. Kemper, *Valuation of Network Effects in Software Markets*,
Contributions to Management Science, DOI 10.1007/978-3-7908-2367-7_18,

of network effects on valuations in less volatile industries. A specialization of the research focus, in turn, could be pursued with respect to specific subsectors of software markets in order to make the models more detailed and realistic. Initial research efforts indicate that network effects can vary significantly across various sectors, but due to the universality of the topological insights on networks it is likely that the developed concepts can be applied to other sectors as well.[1] A particularly interesting research object is the computer game sector due to very strong network effects of computer games and online game platforms. Numerous additional research opportunities derive from the insights of the complex networks analysis of customer networks in software markets. Accordingly, it would be interesting to conduct further research on the identified critical mass of and the derived volatility distribution of cash flows based on the presented investigations of the customer network. Hence, there exist a variety of research opportunities from a investment theoretical perspective based on generalizations or specializations of the research contributions.

18.1.2 Opportunities in Behavioral Finance

Another vital insight of the previous research is that the heterogeneity of consumers has to be considered in order to enhance the understanding of social and economic interactions of individuals. Research on behavioral finance reveals that these factors are also relevant to financial decision making. As behavioral finance pursues complementary research goals, the insights on complex networks research could be used to extend the financial research on bounded rationality. For example, complex networks research could be applied in order to account for the cognitive limitations of economic agents in order to explain the herding behavior of investors. Accordingly, varies additional adoption rules have to be assessed, e.g., the mean-field approximation for large-scale interactions among investors. Thereby, the boundaries of financial research could be extended by accounting for phenomena in relevant large-scale social networks such as investors or shareholders.

18.1.3 Opportunities for Empirical Financial Research

In order to calibrate the parameters of the model for real-world applications in valuations more empirical background research is required. Network metrics, such as connectivity have to be investigated and compared across various valuations. The research revealed that the topologies of social networks has a vital impact on the respective diffusion processes. Therefore, it is necessary to know more about the

[1] Please confer Sect. 5.5.

node distributions of social relationships in the investigated software markets. Such investigations can be performed by surveys, interviews, or a study of other representatives of the social relationships, such as the network of email communication. Specifically designed large-sample studies have to be conducted in order to confirm and to extend the findings. Alternatively, in-depth case studies or longitudinal surveys could contribute further insights. In general, more detailed investigations will enhance the mediation of financial contracting problems in negotiations between corporations and investors due to the design of a systematical approach for the internalization of network effects. An example of existing research in this area are the analysis of ownership networks and management board networks. However, the appropriate relational data and the respective computer programs are required in order to test the network strategies empirically and to enhance the understanding of the unintuitive nature of network externalities. Moreover, a case study could demonstrate the impact from network distributions. In order to derive a proof of concept based on real-world data, it would be necessary to know the actual link distribution, e.g., the German Xing AG, a German business network. Empirical research on the distribution of the customer networks in software markets will provide further research opportunities. In this context it is important to note the findings on the variance and invariance of networks with respect to scaling.

18.2 Complex Networks Research Opportunities

In addition to the financial research opportunities, the book reveals also possibilities to advance network complex networks research. This book is a first roadmap for a variety of unanswered research questions. Its primary focus was to illustrate the importance of network effects and the relevance of complex networks theory in order to approach this analysis. But there remain a variety of research problems, limitations and trade-offs. The most important impulses are depicted in the following.

18.2.1 Heterogeneous Economic Agents

A crucial issue in the cost-benefit consideration of the adoption and diffusion simulator is the quantification of the respective benefits and costs. Although it is plausible to start with the assumption of a standard normal distribution of benefits that averages out in large populations, more specific investigations are required as the model is sensitive to such assumptions. In particular, it is necessary to study the derivative benefits resulting from network effects. Empirical research reveals that for a variety of reasons such investigations are not trivial (Mueller 2003). First, it is necessary to design such research in a manner that the customer has an incentive to state the true benefits, as it is possible to manipulate a statement about the willingness to pay. But, in addition, it is also difficult to quantify the relevant benefits even if the investigated individuals are willing to reveal their actual willingness to pay. Hence, there are a variety of research opportunities which are, however, not in the focus of this book.

18.2.2 Network Generation

As network topologies have implications on the diffusion behavior of networks, the generation of various network types is a decisive step in the framework. For this reason, it is important to investigate systematically the available algorithms for network generation in order to increase the similarity between simulated and real world networks, e.g., simulated annealing. In order to calibrate the generated complex networks model more accurately, it is necessary to know the link distribution in the customer network. Therefore, in order to enhance the precision of the model empirical research is required in order to investigate the connectivity and other network characteristics of the customer network.

18.2.3 Network Evolution

The research model is a static complex networks approach based on a given underlying network structure, but the topology of networks, and in particular social networks, changes over time. Therefore, the evolution of networks is another vital research topic in complex networks research comprising of issues such as the emergence, variation and disappearance of nodes and links within the customer network. Similarly, the process of synchronization is a phenomenon that can be observed in various networks (Strogatz 2004). Hence, it would be interesting to study the synchronization of coexisting networks and the respective interactions which could have an impact on the adoption and diffusion process of both networks.

18.2.4 Nature of Social Networks

All model building activities share that the designer of a partial network analysis is confronted with the dilemma of determining the boundaries of social relations, which are extremely difficult to delineate because of their fuzzy nature. In particular, in partial network analysis, a legitimate network separation is ignored by some network researchers which leads to inaccurate results, since some emerging properties of the overall entity are not identified. Therefore, it is necessary to either make a theoretically informed decision about the significant boundaries of the model based on statistical theory, or to work with the total population. However, real-life networks are generally very large, the complexity of the computational problems that need to be solved in the analysis yield considerable research obstacles that can only be partially overcome by numerical approximations, simplifications, and simulations.

18.2.5 Complex Systems Theory

Finally, it is also possible to identify the following highly relevant chances for interdisciplinary research, in addition to the depicted opportunities for each research discipline. In this context it is important to note that *complex systems theory* is an emerging body of interdisciplinary research that provides a variety of additional concepts and tools that could be explored in order to enhance customer network-centric valuation in software markets (Bar-Yam 1997). According to this research paradigm other dynamical systems, such as cellular automata, exhibit additional insights on networks. For example, cellular automata can be applied in order to study properties of small-world networks or diffusion models such as SIS-models or SIR-models (Bar-Yam 1997; Watts and Strogatz 1998; Watts 1999; Boccara and Fuks 2003; Vazquez et al. 2003; Kemper 2006). The motivation of such studies is either to study opinion formation in social networks, or the effects of network topologies.

18.3 Reconsideration of the Research Outlook

All in all, the research outlook indicates a variety of further insights in order to increase the understanding of counter-intuitive network phenomena. The presented customer network-centric perspective on valuation in software markets may cause additional costs, but it provides additional reliable information and increases the transparency with respect to risks resulting from networks. For this purpose, it provides a battery of statistics, concepts, and models. These can be applied in the presented customer network-centric valuation approach in software market in order to investigate investment decisions. Thereby, the research contributes to the question why management, financial sponsors and financial academics should consider the value of intangible customer networks due to networks effects. It is therefore a starting point to combine the new complex networks perspective with approved financial concepts in order to push the envelope in research on valuation in software markets. The research is based on existing research activities, illustrates the limitations of existing approaches and extends the model in accordance to the identified restrictions. Suggested extensions provide further insights as there remain a variety of questions unsolved for further research projects. Looking forward, we hope that this book contributes to a benevolent integration of complex networks research into various frontiers of management research by increasing its popularity in management theory and practice.

Chapter 19
Executive Summary

> "I will be brief. Not nearly as brief as Salvador Dali, who gave the World's shortest speech. He said, 'I will be brief so brief I have already finished,' and he sat down."
>
> E.O. Wilson, Commencement address at Penn State

The customer base is an important value driver of software companies, and a reliable prediction of its development is of fundamental importance for valuations, investment decisions and strategic management. A particularity in software markets is that an individual's purchasing decision is often influenced by other users' choices, for example the number of communication partners using the same software or the total number of licenses sold. Although the influence of such customer network effects on the diffusion of software products is evident, their quantitative assessment and their impact on the valuation of software companies with conventional approaches remain elusive. This book contributes to closing this gap by developing analytical and numerical methods for measuring network effects and their implications for valuation in software markets.

Actual and potential customers of a software company can be formally modeled as actors in a communication network, e.g., the internet, where a link between two actors implies that these actors influence each other's purchasing decisions. Based on the theory of complex networks, the investigations in this book reveal that the diffusion process in this model highly depends on particular structural properties of the network. The body of research is organized in three main parts: The first part identifies network effects in software market valuations. It reveals the relevance of network effects for valuations in software markets, particularly if the derivative benefits of customer networks are high. In the second part, the focus is moved to the quantification of network effects. An important contribution is the development of the analytical markov matrix diffusion model which provides a description of simple diffusion dynamics in customer networks, if the required input parameters are available. As a main result, this part links valuation to adoption and diffusion models of customer networks in a network effects valuation framework for software markets. The reconsideration of this framework reveals that the disregard of network properties, topologies and dynamics are its main limitations. In the third part, the outlined limitations are approached from a complex networks perspective. Based on a review of the research on properties, topologies and dynamics of complex

A. Kemper, *Valuation of Network Effects in Software Markets*,
Contributions to Management Science, DOI 10.1007/978-3-7908-2367-7_19,

customer networks, a complex networks adoption and diffusion simulation is developed. This simulator is used to derive numerical evidence for the dependence of the diffusion process on particular properties of the underlying customer network. A summary of the findings, a review of important limitations, and an outlook on further research opportunities conclude the book. The interdisciplinary research conducted in this book builds upon observations from social sciences, theoretical physics and graph theory to improve the quality of valuation in software markets by a customer-network centric approach to predicting the development of the customer base. The implications of this research are not restricted to mere financial aspects, but also comprises social and political responsibilities as it aims to prevent irrational corporate failures. Nonetheless, various open questions remain, providing starting points for further research activities. The three most relevant questions are highlighted in the following:

1. *Network generation.* As network topologies have implications on the diffusion behavior of networks, the generation of various network types is a decisive step in the framework. For this reason it is important to investigate if there exist further models for customer networks that capture real world networks more accurately.
2. *Network evolution.* It is important to investigate the evolution of the network of current and potential customers, e.g., emergence, variation and disappearance of nodes and links, on a finer level of detail in the complex networks software market model.
3. *Other industries.* The outlined research and the findings have a focus on software markets. It should be explored whether the outlined methodology can be transferred to other industries.

In summary, the outlined customer network-centric based valuation approach may legitimate additional costs induced by the analysis of customer networks, as it provides a variety of insights that allow enhancing the quality of valuations in software markets.

Bibliography

Achi, Z., Doman, A. Sibony, O. Sinha, J., & Witt, S. (1995). The paradox of fast growth tigers. *McKinsey Quarterly.*

Achleitner, A., & Nathusius, E. (2004). Unternehmensbewertung bei Venture-Capital-Finanzierungen. *Wirtschaftswissenschaftliches Studium, 33*, 134–139.

Adamic, L., & Huberman, B. (2000). Power-law distribution of the world wide web. *Science, 287*, 2115.

Adams, D., Nelson, R. & Todd, P. (1992). Perceived usefulness, ease of use, and usage of information technology: A replication. *Management Information Systems Quarterly, 16*, 227–247.

Ajzen, I., & Fishbein, M. (1980). *Understanding attitudes and predicting social behavior.* Eaglewood Cliffs: Prentice-Hall.

Albert, R. (2001). *Statistical Mechanics of Complex Networks*. Ph. D. book, University of Notre Dame.

Albert, R., & Barabasi, A. (2002). Statistical mechanics of complex networks. *Reviews of Modern Physics, 74*, 47–97.

Albert, R., Jeong, H., & Barabasi, A. (1999). Diameter of the world-wide web. *Nature, 401*, 130–131.

Alfarano, S., & Milakovic, M. (2008). Should network structure matter in agent-based finance? *Economics working papers of Christian-Albrechts-University of Kiel 04*, 1–32.

Amaral, L., Scala, A., Barthélémy, M., & Stanley, H. (2000). Classes of small-world networks. *Proceedings of the National Academy of Sciences of the United States of America, 97*, 11149–11152.

Amoco, B. (2000). The value of real options.

Amram, M., & Kulatilaka, N. (2000). Strategy and shareholder value creation: The real options frontier. *Journal of Applied Corporate Finance, 13*, 8–21.

Anderson, R., & May, R. (1991). *Infectious diseases of humans.* variety of mathematical models for epidemics: Oxford University Press.

Anglin, P., & Gencay, R. (1996). Semiparametric estimation of a hedonic price function. *Journal of Applied Econometrics, 11*(6), 633–648.

Aoki, M. (2005). Modeling a large number of agents by types: Models as large random decomposable structures. In *New Tools of economic dynamics* (pp. 3–24). Berlin: Springer. Agent.Based-Modeling.

Arditti, F. (1996). *Derivatives - A comprehensive resource for options, futures, interest rate swaps and mortgage securities.* Boston: Harvard Business School Press.

Armbrust, M., Fox, A., Griffith, R., Joseph, A., Katz, R., Konwinski, A., Lee, G., Patterson, D., Rabkin, A., Stoica, I., & Zaharia, M. (2009). Above the clouds: A berkeley view of cloud computing. Technical report, University of California at Berkeley.

Armstrong, M. (1998). Network interconnection in telecommunications. *Economic Journal, 108*, 545–564.

Arnold, G., & Hantzopoulos, P. (2000). The theory-practice gap in capital budgeting: Evidence from the united kingdom. *Journal of Business Finance and Accounting, 27*, 603–626.

Arthur, W. (1989). Competing technologies, increasing returns, and lock.in by historical events. *Economic Journal, 99*(394), 116–131.

Arthur, W. (1990). Positive feedbacks in the economy. *Scientific American, 262*, 92–99.

Arthur, W. (1996). Increasing returns and the new world of business. *Harvard Business Review*, 100–109.

Baecker, P., & Hommel, U. (2004). 25 years real options approach to investment valuation: Review and assessment. *Zeitschrift für Betriebswirtschaft Ergänzungsheft, 3*, 1–53.

Bagozzi, R., Davis, F., & Warshaw, P. (1992). Development and test of a theory of technological learning and usage. *Human Relations, 45*(7), 660–686.

Bailey, N. (1975). *The mathematical theory of infectious diseases and its applications*. New York: Hafner Press.

Baldi, S. (1998). *Grundlagen der Wirtschaftsinformatik*. München.

Baldwin, C., & Clarke, K. (2000). *Design rules, Vol. 1: The power of modularity*. Cambridge MA: MIT.

Ballwieser, W. (1987). *Unternehmensbewertung und Komplexitätsreduktion*. Wiesbaden: Gabler.

Ballwieser, W. (1995). Aktuelle Aspekte der Unternehmensbewertung. *Die Wirtschaftsprüfung*, 119–129. objective corporate value.

Bamberg, G., & Coenenberg, A. (2004). *Betriebswirtschaftliche Entscheidungslehre* (12th ed.). München.

Banks, D., & Carley, K. (1996). Models for network evolution. *Journal of Mathematical Sociology, 21*, 173–196.

Bar-Yam, Y. (1997). *Dynamics of complex systems*. Reading MA: Perseus.

Barabasi, A. (2002). *Linked - The new science of networks*. Cambridge: Perseus. Network Theory.

Barabasi, A., & Albert, R. (1999). Emergence of scaling in random networks. *Science, 286*, 509–512.

Barabasi, A., Albert, R., & Jeong, H. (1999). Mean-field theory for scale-free random networks. *Physica A: Statistical Mechanics and its Applications, 272*, 173–187.

Barabasi, A., Albert, R., Jeong, H., & Bianconi, G. (2000). Power-law distribution of the world wide web. *Science, 287*, 2115a.

Barabasi, A., Jeong, H., Ravasz, E., Neda, Z., Schuberts, A., & Vicsek, T. (2002). Evolution of the social network of scientific collaborations. *Physica A: Statistical Mechanics and its Applications, 311*, 590–614.

Barone-Adesi, G., & Whaley, R. (1987). Efficient analytic approximation of american option values. *Journal of Finance, 42*, 301–320.

Bass, F. (1969). A new product growth model for consumer durables. *Management Science, 15*, 215–227.

Bassen, A., & Popovic, T. (2004). Die Bewertung von B2C-E-Commerce-Unternehmen. *Finanz Betrieb*, 838–850.

Bateman, R. (1971). Appraising the going concern. *Appraisal Journal*, 526–534. Comparables Transactions.

Batten, D., Cast, J., & Thord, R. (1995). *Networks in action - communication, economics and human knowledge*. Berlin: Springer. Networks Theory, Networks as Dynamic Systems, Infrastructure Networks, Economic Networks.

Bauer, H., & Hammerschmidt, M. (2005). Customer-based corporate valuation. *Management Decision, 43*, 331–348.

Bayon, T., Gutsche, J., & Bauer H. (2002). Customer equity marketing: touching the intangible. *European Management Journal, 20*(3), 213–222.

Beer, M., & Nohria, N. (2001). *Harvard business review on turnarounds*, Chapter Cracking the code of change, pp. 1–21. Harvard Business Review. Havard Business School Press.

Beinhauer, W., Herr, M., & Schmidt, A. (2008). *SOA für Agile Unternehmen*. Symposion Publishing.

Bellinger, B., & Vahl, G. (1984). *Unternehmensbewertungin Theorie und Praxis*. Wiesbaden: Gabler.

Benaroch, M. (2002). Managing information technology investment risk: A real options perspective. *Journal of Management Information Systems, 19*, 43–84.
Benaroch, M., & Kauffman, R. (1999). A case for using real option pricing analysis to evaluate information technology project investments. *Information Systems Research, 10*, 70–86.
Benaroch, M., & Kauffman, R. (2000). Justifying electronic banking network expansion using real options analysis. *MIS Quarterly, 24*, 197–225.
Berger (2008). Boosting IT efficiency. Roland Berger Strategy Consultants, München.
Berger, A., & Udell, G. (1998). The economics of small business finance: The roles of private equity and debt markets in the financial growth cycle. *Journal of Banking and Finance, 22*, 613–673.
Berger, P., & Nasr, N. (1998). Customer lifetime value: marketing models and its applications. *Journal of Interactive Marketing, 12*(1), 17–30.
Bernoulli, D. (1760). Essai d'une nouvelle analyse de la mortalite causee par la petite verole. *Memoris of Mathematical and Physical Academy of Royal Science Paris,* 1–45. Effect of cowpox inoculation on the spread of smallpox.
Besen, S., & Farrell, J. (1994). Choosing how to compete: Strategies and tactics in standardization. *Journal of Economic Perspectives, 8*(2), 93–115.
Besen, S., & Johnson, L. (1986). *Compatibility standards, competition, and innovation in the broadcasting industry*. Santa Monica: Rand Corporation.
Beutelspacher, A. (2003). *Lineare Algebra*. Vieweg und Teubner.
Bianconi, G., & Barabasi, A. (2001a). BoseŰEinstein condensation in complex networks. *Physical Review Letters, 86*, 5632–5635.
Bianconi, G., & Barabasi, A. (2001b). Competition and multiscaling in evolving networks. *Europhysics Letters, 54*, 436–442.
Black, F., & Scholes, M. (1973). The pricing of options and corporate liabilites. *Journal of Political Economy, 81*, 637–654.
Blanchard, P., Chang, C., & Krueger, T. (2002). Epidemic thresholds on scale-free graphs: The interplay between exponent and preferential choice. *Preprint cond-mat/0207319*, 0207319.
Blattberg, R., & Deighton, J. (1996). Manage marketing by the customer equity test. *Harvard Business Review, 74*, 136–144.
Blattberg, R., Getz, G., & Thomas, J. (2001). *Customer equity: Building and managing relationships as valuable assets*. Cambridge, MA: Harvard Business School Press.
Blattberg, R., & Thomas, J. (1998). The Fundamentals of Customer Equity Managment. In *Handbuch Kundenbindungsmanagement* (pp. 329–357). Wiesbaden: Gabler Verlag.
Blomeyer, E. (1986). An analytic approximation for the american put price for options on stock with dividends. *Journal of Financial and Quantitative Analysis, 21*, 229–233.
Blumling, M., Frick, K., & Meehan, W. (2002). A hard turnaround for software. *The McKinsey Quarterly*.
Boccara, N., & Cheong, K. (1992). Automata network sir models for the spread of infectious diseases in populations of moving individuals. *Journal of Physics. A: Mathematical General, 25*, 2447–2461.
Boccara, N., & Fuks, H. (2003). Modeling diffusion dynamics with probablistic cellular automata. *arXiv:adap-org/9805004v1*.
Bockemuehl, M. (2001). *Realoptionspreistheorie und die Bewertung von Produktinnovationen: Der Einfluss von Wettbewerbseffekten*. Wiesbaden: Gabler DUV.
Boer, P. (2002). *The Real Options Solution: Finding Value in a High-Risk World*. New York: Wiley.
Boguna, M., Pastor-Satorras, R., & Vespignani, A. (2002). Absence of epidemic threshold in scale-free networks with connectivity correlations. *Preprint cond-mat/0208163*, 0208163.
Bollobas, B. (1985). *Random Graphs*. London: Academic.
Bollobas, B. (2001). *Random Graphs* (2nd ed.). New York: Academic.
Bollobas, B., & Riordan, O. (2006). *Percolation*. Cambrige/MA: Cambridge University Press.
Booch, G. (2007). *Object-oriented Analysis and Design with Applications*. New York: Addison-Wesley.

Born, K. (1995). *Unternehmensanalyse und Unternehmensbewertung*. Stuttgart: Schäffer-Poeschel Verlag. Definition of Corporate Value (pp. 21–22).

Bower, J., & Christensen, C. (1995). Disruptive technologies: Catching the wave. *Harvard Business Review, Jan-Feb*, 43–53.

Box, J. (1983). Extending product lifetime: Prospects and opportunities. *European Journal of Marketing, 17*, 34–49.

Boyle, P. (1977). Options: a monte carlo approach. *Journal of Financial Economics, 4*, 323–338.

Brach, M. (2003). *Real options in practice*. Hoboken, NJ: John Wiley & Sons.

Brealey, R., & Myers, S. (1996). *Principles of corporate finance*. New York: McGraw-Hill.

Brennan, M., & Schwartz, E. (1978). Finite difference methods and jump processes arising in the pricing of contingent claims: a synthesis. *Journal of Financial and Quantitative Analysis, 13*, 462–474.

Broadbent, S., & Hammersley, J. (1957). Percolation processes: I. crystals and mazes. *Proceedings of the Cambridge Philosophical Society, 53*, 629–641.

Brockmann, D., Hufnagel, L., & Geisel, T. (2006). The scaling laws of human travel. *Nature, 439*, 462–465.

Broder, A., Kumar, R., Maghoul, F., Raghavan, P., Ra-jagopalan, S., Stata, R., Tomkins, A., & Wiener, J. (2000). Graph structure in the web. *Computer Networks, 33*, 309–320.

Brynjolfsson, E. (1996). The contribution of information technology to consumer welfare. *Information Systems Research, 7*, 281–300.

BSA (2007). Fourth annual bsa and idc global software piracy study 2006. www.bsa.org/globalstudy.

Buchanan, M. (2002). *Nexus: Small worlds and the groundbreaking science of networks*. New York: W.W. Norton & Company Inc.

Bulow, J., Geanakoplos, J., & Klemperer, P. (1985). Multimarket oligopoly: Strategic substitutes and complements. *Journal of Political Economy, 93*, 488–511.

Busse von Colbe, W. (1957). *Der Zukunftserfolg: Die Ermittlung des künftigen Unternehmenserfolges und seiner Bedeutung für die Bewertung von Industrieunternehmen*. Wiesbaden.

Busse von Colbe, W., & Coenenberg, A. (1992). *Unternehmensakquisition und Unternehmensbewertung: Grundlagen und Fallstudien*. Stuttgart: Schäffer-Poeschel.

Buxmann, P. (1996). *Standardisierung betrieblicher Informationssysteme*. Wiesbaden.

Buxmann, P. (2001). *Informationsmanagement in vernetzten Unternehmen*. Wiesbaden: Gabler DUV.

Buxmann, P., & König, W. (1998). Das Standardisierungsproblem: Ein ökonomisches Entscheidungsproblem zur Auswahl von Standards in betrieblichen Informationssystemen. *Wirtschaftsinformatik, 40*, 122–129.

Buxmann, P., Weitzel, T., & König, W. (1999). Auswirkung alternativer Koordinationsmechanismen auf die Auswahl von Kommunikationsstandards. *Zeitschrift für Betriebswirtschaft, 2*, 133–151.

Callaway, D., Hopcroft, J., Kleinberg, J. Newman, M., & Strogatz, S. (2001). Are randomly grown graphs really random? *Physical Review E, 64*, 041902.

Camerer, C., Loewenstein, G., & Prelec, D. (2005). Neuroeconomics: How neuroscience can inform economics. *Journal of Economic Literature, 43*, 9–64.

Carter, M., & Wright, J. (1999). Interconnection in network industries. *Review of Industrial Organization, 14*, 1–25.

Cauchy, A. (1813). Recherche sur les polyèdres - premier mémoire. *Journal de l'Ecole Polytechnique 9, Cahier 16*, 66–86.

Cayley, A. (1875). Über die Analytischen Figuren, welche in der Mathematik Bäume genannt werden und ihre Anwendung auf die Theorie chemischer Verbindungen. *Berichte der deutschen Chemischen Gesellschaft, 8*, 1056–1059.

Chalasani, P., Jha, S., & Sullivan, K. (1997). *The options approach to software prototyping decisions*. Carnegies Mellon University.

Choi, J. (1996). Network externality, compatibility choice, and planned obsolescence. *The Journal of Industrial Economics, 42*(2), 167–182.

Choi, J., Kristiansen, E., & Nahm, J. (2005). Strategic product pre-announcements in markets with network effects. *NET Institute, 9*, Working Paper 05–09.

Choi, S., & Winston, A. (2000). Benefits and requirements for interoperability in electronic marketplace. *Technology in Society, 22*(1), 33–44.

Chou, D., & Shy, O. (1990). Network effects without network externalities. *International Journal of Industrial Organization, 8*, 259–270.

Citigroup (2006). Software & IT services: An industry in transition. Technical report, Citigroup.

Clement, M., Litfin, T., & Peters, K. (1998). *Marketing mit interaktiven Medien Ǘ Strategien zum Markterfolg*, Chapter Netzeffekte und kritische Masse, pp. 81–94. Frankfurt a. M.

Clemons, E. (1991). Evaluation of strategic investments in information technology volume. *Communications of the ACM, 34*, 22–36.

Cohen, R., ben Avraham, D., & Havlin, S. (2002). Efficient immunization of populations and computers. *cond-mat/0207387*.

Cohen, R., Erez, K., ben Avraham, D., & Havlin, S. (2000). Resilience of the internet to random breakdowns. *Physical Review Letters, 85*, 4626–4628.

Cohen, R., Erez, K., ben Avraham, D., & Havlin, S. (2001). Breakdown of the internet under intentional attack. *Physical Review Letters, 86*, 3682–3685.

Coleman, J., Katz, E., & Menzel, H. (1957). The diffusion of an innovation among physicians. *Sociometry, 20*(4), 253–270.

comScore (2008). Social networking goes global.

Copeland, T., & Antikarov, V. (2001). *Real Options: A Practioner's Guide*. New York: Texere LLC.

Copeland, T., Koller, T., & Murrin, J. (1996). *Valuation: measuring and managing the value of companies*. New York:Wiley.

Copeland, T., & Weston, J. (1988). *Financial theory and corporate policy*. Reading, MA: Addison-Wesley Series in Finance.

Cornelsen, J. (2002). What is the customer worth? Customer value analysis in relationship marketing on the basis of car buyers. *Yearbook of Marketing and Consumer Research, 1*, 27–46.

Cox, J., & Ross, S. (1976). The valuation of options for alternative stochastic processes. *Journal of Financial Economics, 3*, 145–166.

Cox, J., Ross, S., & Rubinstein, M. (1979). Option pricing: A simplified approach. *Journal of Financial Economics, 7*, 229–263.

Cox, J., & Rubinstein, M. (1985). *Option markets*. New Jersey: Englewood Cliffs.

Coy, P. (1999). Exploiting uncertainty - the real options revolution in decision making. *Business Week*, 118–124.

Damodaran, A. (1996). *Investment valuation*. New York: Wiley.

Damodaran, A. (2001). *The dark side of valuation: Firms with no earnings, no history and no comparables*.

D'Aveni, R. (1994). *Hypercompetition: Managing the dynamics of strategic maneuvering*. New York: Free Press.

David, P. (1985). Clio and the economies of qwerty. *American Economic Review, 75*(2), 332–337.

David, P., & Bunn, J. (1988). The economics of gateway technologies and network evolution: Lessons from electricity supply history. *Information Economics and Policy, 3*, 165–202.

David, P., & Greenstein, S. (1990). The economics of compatibility standards: an introduction to recent research. *Economics of Innovation and New Technology, 1*(1), 3–41.

Davis, F., Bagozzi, R., & Warshew, P. (1989). User acceptance of computer technology: a comparison of two theoretical models. *Management Science, 35(8)*, 982–1003.

Davis, G. (1998). Estimating volatility and dividend yield when valuing real options to invest or abandon. *The Quarterly Review of Economics and Finance, 38*, 725–754.

Davis, G., & Greve, H. (1997). Corporate elite networks and governance changes in the 1980s. *American Journal of Sociology, 103*, 1–37.

Day, G. (1981). The product life cycle: Analysis and applications issues. *Journal of Marketing, 45*, 60–67.

Day, G., & Fahey, L. (1988). Valuing marketing strategies. *Journal of Marketing, 52*(3), 45–57.

DeBondt, W. (1995). *Behavioral Finance and Decision Theory in Investment Management*, Chapter Investor Psychology and the Dynamics of Security Prices.

Dessein, W. (2001). *Network competition in nonlinear pricing*. Chicago: University of Chicago.

Dhalla, N., & Yuspeh, S. (1976). Forget the product life cycle concept. *Harvard Business Review, 54*, 102–112.

Dixit, A., & Pindyck, R. (1994). *Investment under uncertainty*. Princeton, NJ: Princeton University Press.

Dodds, P., Muhamad, R., & Watts, D. (2003). An experimental study of search in global social networks. *Science, 301*, 827–829.

Dorogovtsev, S., & Mendes, J. (2001). Effect of the accelerating growth of communications networks on their structure. *Physical Review E, 63*, 025101.

Dorogovtsev, S., & Mendes, J. (2002). Evolution of networks. *Advances in Physics, 51*, 1079–1187.

Dorogovtsev, S., Mendes, J., & Samukhin, A. (2001). Giant strongly connected component of directed networks. *Physical Review E, 64*, 025101.

DosSantos, B. (1991). Justifying investment in new information technologie. *Journal of Management Information Systems, 7*, 71–89.

Doyle, P. (2000). Value-based marketing. *Journal of Strategic Marketing, 8*(4), 299–311.

Duden. (1989). *Duden Etymologie: Herkunftswörterbuch der deutschen Sprache* (2nd ed.). Mannheim: Duden.

Dwyer, F. (1997). Customer lifetime valuation to support marketing decision making. *Journal of Direct Marketing, 4 (Fall)*, 8–15.

Dybvig, P., & Spatt, C. (1983). Adoption externalities as public goods. *Journal of Public Economics, 20*, 231–247.

Ebel, H., Mielsch, L., & Bornholdt, S. S. (2002). Scale-free topology of e-mail networks. *Physical Review E, 66*, 035103.

Economides, N. (1996). The economics of networks. *International Journal of Industrial Organization, 14*(6), 673–699.

Economides, N., & Himmelberg, C. (1995). Critical mass and network evolution in telecommunications In *Toward a competitive telecommunications industry*. Selected Papers of Telecommunications Policy Research Conference 1994.

Economides, N., & White, L. (1993). One-way networks, two-way networks, compatibility, and antitrust. *Department of Economics Working Paper, New York University ED-93-14*, 1331–1355.

Eguiluz, V., & Klemm, K. (2002). Epidemic threshold in structured scale-free networks. *Physical Review Letters, 89*, 108701.

EITO. (2006). *European information technology observatory yearbook 2006*. European Information Technology Observatory.

Erdos, P., & Renyi, A. (1959). On random graphs. *Publicationes Mathematicae, 6*, 290–297.

Evans, P., & Wurster, T. (2000). *Blown to bits: How the new economics of information transforms strategy*. Boston, MA: Harvard Business School Press.

Everitt, B. (1974). *Cluster analysis*. New York: John Wiley.

Faloutsos, M., Faloutsos, P., & Faloutsos, C. (1999). On power-law relationships of the internet topology. *Computer Communications Review, 29*, 251–262.

Fama, E. (1970). Efficient capital markets. *Journal of Finance, 25*, 383–417.

Farrell, J., & Klemperer, P. (2001). *Coordination and Lock-In: Competition with Switching Costs and Network Effects*.

Farrell, J., & Saloner, G. (1985). Standardization, compatibility, and innovation. *Rand Journal of Economics, 16*, 70–83.

Farrell, J., & Saloner, G. (1986). Installed base and compatibility: Innovation, product preannouncement, and predation. *American Economic Review, 76*(4), 940–955.

Farrell, J., & Saloner, G. (1987). Competition, Compatibility and Standards: The economics of horses, penguins and lemmings. In *Product standardization and competitive strategy* (pp. 1–21). Amsterdam: North Holland.
Fell, D., & Wagner, A. (2000). The small world of metabolism. *Nature Biotechnology, 18*, 1121–1122.
Fischer, S. (1978). Call option pricing when the exercise price is uncertain, and the valuation of index bonds. *Journal of Finance, 33*, 169–176.
Freeman, L. (1977). A set of measures of centrality based upon betweenness. *Sociometry, 40*, 35–41.
Fronczak, A., Holyst, J., Jedynak, M., & Sienkiewicz, J. (2002). Higher order clustering coefficients in barabasi-albert networks. *Physica A: Statistical Mechanics and its Applications, 316*, 688–694.
Fulgoni, G. (2007). Consumer trends in social networking. *ComScore*, http://www.comscore.com/blog/2007/10/consumer_trends_in_social_netw.html, 21.1.2008.
Galaskiewicz, J., & Marsden, P. (1978). Interorganizational resource networks: Formal patterns of overlap. *Social Science Research, 7*, 89–107.
Gamma, E., Helm, R., & Johnson, R. (1995). *Design Patterns. Elements of Reusable Object-Oriented Software*. Amsterdam: Addison-Wesley Longman.
Gandal, N. (1994). Hedonic price indeces for spreadsheets and an empirical test of the network externalities hypothesis. *Rand Journal of Economics, 25*(1), 160–170.
Gareis, R. (2000). Old boys' network. *Data Communications*.
Garfield, E. (1979). It's a small world after all. *Current Contents, 43*, 5–10.
Gerhardt, T. (1992). *Strategie und Struktur in der deutschen Softwareindustrie. Eine industrieökonomische Untersuchung der Unternehmensentwicklung in der Softwarebranche*. München: VVF-Verlag.
Geske, R. (1978). The valuation of compound options. *Journal of Financial Economics, 7*, 63–81.
Geske, R., & Johnson, H. (1984). The american put option valued analytically. *Journal of Finance, 39*, 1511–1524.
Geske, R., & Shastri, K. (1985). Valuation by approximation: a comparison of alternative option valuation techniques. *Journal of Financial and Quantitative Analysis, 20*, 45–71.
Goldenberg, D. (1991). A unified method for pricing options on diffusion processes. *Journal of Financial Economics, 29*, 3–34.
Goldstein, S. (1998). *The Role of Brands in the Digital Economy*. AT Kearney.
Gooch, L., & Grabowski, J. (1976). Advanced valuation methods. In *Mergers and acquisitions* (pp. 15–29).
Gordon, M. (1959). Dividends, earnings and stock prices. *Review of Economics and Statistics, 41*, 99–105.
Graham, J., & Harvey, C. (2001). The theory and practice of corporate finance: Evidence from the field. *Journal of Financial Economics, 60*, 187–243.
Granovetter, M. (1978). Threshold models of collective behavior. *American Journal of Sociology, 83*(6), 1420–1443.
Grantham, L. (1997). The validity of the product life cycle in the high-tech industry. *Marketing Intelligence & Planning, 15*, 4–10.
Grassberger, P. (1983). On the critical behavior of the general epidemic process and dynamical percolation. *Mathematical Biosciences, 63*, 157–172.
Graumann, M. (1993). Die Ökonomie von Netzprodukten. *Zeitschrift für Betriebswirtschaft, 63*, 1331–1355.
Grochla, E. (1978). *Einführung in die Organisationstheorie*. Stuttgart: Schaeffer-Poeschel.
Groeger, L. (2008). *Soziale Epidemien: Das Phänomen exponentieller Produktverbreitung*. Fördergesellschaft Produkt-Marketing E.V., Köln.
Groehn, A. (1996). *Netzeffekte in der Software-Industrie: Eine Analyse der empirischen Literatur*. Number 743. Kiel: The Kiel Institute of World Economics. Network Effects in Software Markets.

Groehn, A. (1997). *Ein Modell der Netzeffekte in der Software-Industrie*. Number 790. Kiel: The Kiel Institute of World Economics. Network Effects in Software Markets.

Groehn, A. (1999). *Netzeffekte und Wettbewerbspolitik - Eine ökonomische Analyse der Softwaremarktes*. Number 296. Tübingen: Mohr Siebeck, Tübingen. Network Effects in Software Markets.

Grossman, G., & Helpman, E. (1994). Endogenous innovation in the theory of growth. *Journal of Economic Perspectives, 8*, 23–44.

Guare, J. (1990). *Six degrees of separation: A play*. New York: Vintage.

Gupta, S., Lehmann, D., & Stuart, J. (2001). *Valuing Customers*, Volume Working Paper No. 02-119. Marketing Science Institute, Cambridge.

Hamilton, A. (2000). Gold delta hedge trap (part 2). Goldeagle, http://www.zealllc.com/2000/delta2.htm.

Harary, F., Morman, R., & Cartwright, D. (1965). *Structural Models: An Introduction to the Theory of Directed Graphs*. New York: John Wiley & Sons.

Hartmann, R., & Teece, D. (1990). Product emulation strategies in the presence of reputation effects and network externalities: some vidence from the microcomputer industry. *Economics of innovation and new technology, 1–2*, 157–182.

Hayek, F. (1937). Economics and knowledge. *Economica, 4*, 33–54.

Hayes, B. (2000a). Graph theory in practice: Part i. *American Scientist, 88*, 9–13.

Hayes, B. (2000b). Graph theory in practice: Part ii. *American Scientist, 88*, 104–109.

Hecker, M. (1998). *Informationsüberflutung und deren Vermeidung*. Hamburg.

Heise (2008). Superrechner Jugene und Ranger nehmen den Betrieb auf. *Heise Online 22.02.2008 16:45*.

Heisenberg, W. (1967). Quantum Mechanics In *Sources of quantum mechanics*. New York: Dover.

Hensel, M., & Wirsam, J. (2008). *Diffusion von Innovationen: Das Beispiel Voice over IP*. Wiesbaden: Gabler DUV.

Herget, M. (1987). *Technical Standards and Competition in the Microcomputer Industry*.

Hertz, D. (1964). Risk analysis in capital investment. *Harvard Business Review, 42*, 95–106. Simulation Process of Numerical Risk Analysis.

Hess, T. (2000). Netzeffekte: Verändern neue Informations- und Kommunikationstechnologien das klassische Marktmodell? *Wirtschaftswissenschaftliches Studium, 29*, 96–98.

Hethcote, H. (2000). Mathematics of infectious diseases. *SIAM Review, 42*, 599–653.

Ho, S., & Pike, R. (1991). Risk analysis in capital budgeting contexts: simple or sophisticated. *Accounting and Business Research, 21*(3), 227–38.

Hoch, D., Roeding, C., Pukert, G., & Lindner, S. (1999). *Secrets of software success*. Boston, MA: Harvard Business School Press.

Hodgson, G. (1993). *The Economics of institutions*. Hants: Edward Elgar Publishing.

Hoekstra, J., & Huizingh, E. (1999). The lifetime value concept in customer-based marketing. *Journal of Market Focused Management, 3*, 257–274.

Hogan, J., Lehman, D., Merino, M., Srivastava, R., Thomas, J., & Verhoef, P. (2002). Linking customer assets to financial performance. *Journal of Service Research, 5*(1), 26–38.

Holme, P., Kim, B., Yoon, C., & Han, S. (2002). Attack vulnerability of complex networks. *Physical Review E, 65*, 056109.

Hommel, U. (1999). Der Realoptionsansatz: Das neue Standardverfahren der Investitionsrechnung. *M & A Review, 1*, 22–29.

Hommel, U., & Kemper, A. (2006). *Handbuch Unternehmensrestrukturierung*, Chapter Sicherung der Netzwerkintegrität in der Unternehmenssanierung, pp. 625–644. Gabler.

Hommel, U., & Müller, J. (1999). Realopionsbasierte Investitionsbewertung. *Finanz-Betrieb, 1*, 177–188.

Hommel, U., & Pritsch, G. (1999). Marktorientierte Investitionsbewertung mit dem Realoptionsansatz. *Finanzmarkt- und Portfoliomanagement, 13*, 121–144.

Hommel, U., Scholich, M., & Baecker, P. (2003). *Reale Optionen: Konzepte, Praxis und Perspektiven strategischer Unternehmensfinanzierung*. Berlin: Springer.

Hufnagel, L., Brockmann, D., & Geisel, T. (2004). Forecast and control of epidemics in a globalized world. *PNAS, 101*, 15124–15129. Global epidemics, SARS, Spread of SARS with Network Model, Stochastic-deterministic model.

Huillier, S. (1861). Mémoire sur la polyèdrométrie. *Annales de Mathématiques, 3*, 169Ű189.

Hull, J. (1989). *Options, futures, and other derivatives securities*. Englewood Cliffs, NJ: Prentice-Hall.

Hull, J., & White, A. (1994). Numerical procedures for implementing term structure modelsi: Single-factor models. *Journal of Derivatives 2*, 7–16.

Ichbiah, D. (1993). *The Making of Microsoft*. Rocklin: Prima Publishing.

Ito, K. (1951). On stochastic differential equations. *Memoirs of the American Mathematical Society, 4*, 1–51.

Jackson, D. (1992). In quest of the grail: breaking the barriers to customer valuation. *Direct Marketing, 55*, 44–47.

Jain, D., & Singh, S. (2002). Customer lifetime value research in marketing: a review and future directions. *Journal of Interactive Marketing, 16*(2), 34–46.

Jansen, D. (1999). *Einführung in die Netzwerkanalyse. Grundlagen, Methoden, Anwendungen.* Opladen: Leske & Budrich.

Janson, S., Luczak, T., & Rucinski, A. (1999). *Random Graphs*. New York: John Wiley.

Jarrow, R., & Rudd, A. (1983). *Option Pricing*. Homewood II: Irwin.

Jeong, H., Mason, S., Barabasi, A., & Oltvai, Y. (2001). Lethality and centrality in protein networks. *Nature, 411*, 41–42.

Jeong, H., Tombor, B., Albert, R., Oltvai, Z., & Barabasi, A. (2000). The large-scale organization of metabolic networks. *Nature, 407*, 651–654.

Jones, S. (1992). Was there a hawthorne effect? *American Journal of Sociology, 98*, 451–468.

Jones, T., & Sasser, W. (1995). Why satisfied customers defect. *Harvard Business Review, Cambridge, 73*, 88–89.

Kahneman, D., Knetsch, J., & Thaler, R. (1990). Experimental tests of the endowment effect and the coase theorem. *The Journal of Political Economy, 98*(6), 1325–1348.

Kahneman, D., & Tversky, A. (1979). A prospect theory: An analysis of decision under risk. *Econometrica, 47*, 263–291.

Kalkota, R., & Robinson, M. (2001). *E-business 2.0: Roadmap for success*. Reading, MA: Addison-Wesley.

Kambil, A., Henderson, C., & Mohsenzadeh, H. (1993). *Strategic Information Technology Management: Perspectives on Organizational Growth and Competitive Advantage*, Chapter Strategic Management of Information Technology: An Options Perspective. Middletown, PA: Idea Publishing Group.

Karonski, M. (1982). A review of random graphs. *Journal of Graph Theory, 6*, 349–389.

Katz, M., & Shapiro, C. (1985). Network externalities, competition, and compatibility. *American Economic Review, 75*(3), 424–440.

Katz, M., & Shapiro, C. (1986). Technology adoption in the presence of network externalities. *Journal of Political Economy, 94*(2), 822–841.

Katz, M., & Shapiro, C. (1992). Product introduction with network externalities. *Journal of Industrial Economics, 40*(1), 55–83.

Katz, M., & Shapiro, C. (1994). Systems competition and network effects. *Journal of Economic Prespctives, 8*, 93–115.

Kauffman, R., & Xiaotong, L. (2005). Technology competition and optimal investment timing: a real options perspective. *IEEE Transactions on Engineering Management, 52*, 15–29.

Keiber, K., Kronimus, A., & Rudolf, M. (2002). Bewertung von Wachstumsunternehmen am Neuen Markt. *Zeitschrift für Betriebswirtschaft, 72*, 735–764.

Kellner, C. (2008). *Diffusion innovativer Technologien am Beispiel brennstoffzellengetriebener PKWs*. Berlin: Logos.

Kellogg, D., & Charnes, J. (2000). Real-option valuation for a biotechnology company. *Financial Analysts Journal, 56*(3), 76–84.

Kemper, A. (2006). Modelling network growth with network theory and cellular automata. *Journal of Modelling in Management, 1*, 75–84.

Kester, W. (1984). Today's options for tomorrow's growth. *Harvard Business Review, 62*(2), 153–160.

Kühn, R., Fuhrer, U., & Jenner, T. (2000). Reale Optionen. *Die Unternehmung, 54*(1), 43–56.

Kühnemann, M. (1985). *Objektorientierte Unternehmensbewertung*. Bern.

Kim, B., & Kim, S. (1999). Measuring upselling potential for life insurance customers. *Journal of Interactive Marketing, 13*(4), 2–9.

Kirchhoff, G. (1847). Über die Auflösung der Gleichungen, auf welche man bei der Untersuchung der linearen Verteilung galvanischer Ströme geführt wird. *Annals of Physical Chemistry, 72*, 497–508.

Kirsch, W. (1984). *Wissenschaftliche Unternehmensführung oder Freiheit vor der Wissenschaft?* Herrsching, Wartaweil.

Klein, B., Alchian, A., & Crawford, R. (1978). Vertical integration, approprable rents, and the competitive contracting process. *Journal of Law and Economics, 21*, 297–326.

Klovdahl, A., Potterat, J., Woodhouse, D., Muth, J., Muth, S., & Darrow, W. (1994). Social networks and infectious disease: The colorado springs study. *Social Science & Medicine, 38*, 79–88.

Koch, C. (2000). *Optionsbasierte Unternehmensbewertung - Realoptionen im Rahmen von Akquisitionen*. Wiesbaden: Gabler DUV.

Kogut, B., & Walker, G. (2003). Restructuring or disintegration of the german corporate network: Globalization as a fifth column. *William Davidson Institute, Working Paper 591*.

Koller, T., Goedhart, M., & Wessels, D. (2005). *Valuation: Measuring and Managing the Value of Companies*. New York: Wiley.

Kollmann, T. (1998). *Die Akzeptanz technologischer Innovationen - eine absatztheoretische Fundierung am Beispiel von Multimedia-Systemen*. University Trier.

Komlós, J., Shokoufandeh, A., Simonovits, M., & Szemerédi, E. (2002). *The regularity lemma and its applications in graph theory*. New York: Springer.

Koppelmann, U. (2006). *Marketing: Einführung in Entscheidunggsprobleme des Absatzes und der Beschaffung*. Stuttgart: Lucius & Lucius.

Korb, J. (2000). *Kaufprozesse im Electronic Commerce: Einflüsse veränderter Kundenbedürfnisse auf die Gestaltung*. Wiesbaden.

Kotler, P., & Bliemel, F. (1999). *Marketing-Management: Analyse, Planung, Umsetzung und Steuerung*. Schaeffer-Poeschel.

Krafft, M., Rudolf, M., & Rudolf-Sipötz, R. (2002). Valuation of customers in growth companies - a scenario based model. *Scientific Working Paper Series of the Center for Market-oriented Management, 25*, 1–22.

Krapivsky, P., Rodgers, G., & Redner, S. (2001). Degree distributions of growing networks. *Physical Review Letters, 86*, 5401–5404.

Kretzschmar, M., Duynhoven, Y., & Severijnen, A. (1996). Modeling prevention strategies for gonorrhea and chlamydia using stochastic network simulations. *American Journal of Epidemiology 114*, 306–317.

Kruschwitz, L. (2005). *Investitionsrechnung* (10th ed.). München: Oldenbourg. Criticism of Real Options: Illiquidity of underlying asset, Indivisibility of Real Investments.

Köster, D. (1998). *Wettbewerb in Netzproduktmärkten*. Wiesbaden: Gabler DUV.

Kumar, R. (1996). A note on project risk and option value of investments in information technologies. *Journal of Management Information Systems, 13*, 187–193.

Laffont, J., Rey, P., & Tirole, J. (1998a). Network competition I: Overview and nondiscriminatory pricing. *Rand Journal of Economics, 29*, 1–37.

Laffont, J., Rey, P., & Tirole, J. (1998b). Network competition II: Price discrimination. *Rand Journal of Economics, 29*, 38–56.

Laffont, J., & Tirole, J. (2000). *Competition in Telecommunications*. Cambridge, MA: MIT.

Lander, D., & Pinches, G. (1998). Challenges to the practical implementation of modeling and valuing real options. *The Quarterly Review of Economics and Finance, 38*, 537–567.

Laux, H. (2005). *Entscheidungstheorie* (6th ed.). Berlin: Springer.

Leibenstein, H. (1950). Bandwagon, snob, and veblen effects in the theory of consumerst' demand. *The Quarterly Journal of Economics, 64*, 183–207.

Leskovec, J., & Horvitz, E. (2008). Planetary-scale views on an instant-messaging network. *arXiv:0803.0939v1*.

Leslie, K., & Michaels, M. (1997). The real power of real options. *The McKinsey Quarterly, 3*, 5–22.

Lev, B. (2001). *Intangibles - Management, Measurement, and Reporting*. Washington, D.C.: Brookings Institution.

Levett, P., Page, M., Nel, D., Pitt, L., Berthon, P., & Money, A. (1999). Towards an application of option pricing theory in the valuation of customer relationships. *Journal of Strategic Marketing, 7*, 275–284.

Levinson, J. (1984). *Guerilla Market Secrets for Making Big Profits from Your Small Business*. Boston: Houghton Mifflin.

Levitt, T. (1965). Exploit the product life cycle. *Harvard Business Review, 43*, 81–94.

Libai, B., Narayandas, D., & Humby, C. (2002). Toward an individual customer profitability model. *Journal of Service Research, 5*, 69–76.

Liebowitz, S., & Margolis, S. (1990). The fable of the keys. *Journal of Law and Economics, 33*, 1–25.

Liebowitz, S., & Margolis, S. (1994). Network externality: An uncommon tragedy. *The Journal of Economic Perspectives, 8*(2), 133–150.

Liebowitz, S., & Margolis, S. (1999). Winners, losers, and microsoft: Competition and antitrust in high technology. *The Independent Institute*.

Lilien, G., Kotler, P., & Moorthy, S. (1992). *Marketing Models*. New Jersey: Prentice-Hall.

Liljeros, F., Edling, C., Amaral, L., Stanley, H., & Aberg, Y. (2001). The web of human sexual contacts. *Nature, 411*, 907–908.

Lintner, J. (1965). The valuation of risk assets and the selection of risky investments in stock portfolios and capital budgets. *Review of Economics and Statistics, 47*, 13–37.

Lucke, C. (2001). *Investitionsprojekte mit mehreren Realoptionen - Bewertung und Analyse*. Wissenschaft und Praxis.

Luehrmann, T. (1998). Investment opportunities as real options: Getting started on the numbers. *Harvard Business Review, 76*, 51–61.

Luterotti, E., & Stefanelli, U. (2002). Existence result for the one-dimensional full model of phase transitions. *Journal for Analysis and its Applications, 21*, 335–350.

Maaß, C., & Pietsch, G. (2008). Web 2.0 auf dem Prüfstand: Zur Bewertung von Internet-Unternehmen. *FernUniversität in Hagen, 431*, 1–24.

Mahajan, V., Muller, E., & Bass, F. (1990). New product diffusion models in marketing: A review and directions for research. *Journal of Marketing, 54*(1), 1–26.

Mahajan, V., & Petersonf, A. (1985). *Models for Innovation Diffusion*. Beverly Hills, CA: Sage Publications.

Mankiw, N. (2004). *Grundzüge der Volkswirtschaftslehre* (3rd ed.). Stuttgart: Schaeffer-Poeschel.

Mansfield, E. (1961). Technical change and the rate of imitation. *Econometrica, 29*, 741–766. Diffusion Theory.

Margrabe, W. (1978). The valuation of an option to exchange one asset for another. *Journal of Finance, 33*, 177–186.

Mariolis, P. (1975). Interlocking directorates and control of cor- porations: The theory of bank control. *Social Science Quarterly, 56*, 425–439.

Marwell, G., Oliver, P., & Prahl, R. (1988). Social networks and collective action: A theory of the critical mass. *American Journal of Sociology, 94*, 503–534.

Maslov, S., & Sneppen, K. (2002). Specificity and stability in topology of protein networks. *Science, 296*, 910–913.

Matutes, C., & Regibeau, P. (1988). Mix and match: Product compatibility without network externalities. *The RAND Journal of Economics, 19*(2), 221–234.

Matutes, C., & Regibeau, P. (1992). Compatibility and bundling of complementary goods in a duopoly. *The Journal of Industrial Economics, 40*(1), 37–54.

Matutes, C., & Regibeau, P. (1996). A selective review of the economics of standardization. entry deterrence, technological progress and international competition. *European Journal of Political Economy, 12*(2), 183–209. first Source of Diffusion Research.

Mauboussin, M., Schay, A., & Kawaja, S. (2000). *Network to Net Worth*. Credit Suisse First Boston. New York: CSFB Equity Research.

May, R., & Anderson, R. (1988). The transmission dynamics of human immunodeficiency virus (hiv). *Philosphical Transactions of the Royal Society of London B, 321*, 565Ű607.

McDonald, R., & Siegel, D. (1986). The value of waiting to invest. *The Quarterly Journal of Economics, 4*, 707–728.

McKendrick, A. (1926). Applications of mathematics to medical problems. *Proceedings of the Edinburgh Mathematical Society, 44*, 98–130.

Meffert, H. (2000). *Marketing*. Wiesbaden: Gabler.

Meise, F. (1998). *Realoptionen als Investitionskalkül: Bewertung von Investitionen unter Unsicherheit*. München: Oldenbourg.

Meissner, W. (1970). Zur Methodologie der Simulation. *Zeitschrift für die gesamte Staatswissenschaft, 126*, 385–402.

Merton, R. (1968). The matthew effect in science. *Science, 159*, 56–63.

Merton, R. (1976). Option pricing when underlying stock returns are discontinuous. *Journal of Financial Economics, 3*, 125–144.

Merton, R. (1995). *Continuous-time finance*. Cambridge, MA: Blackwell, Oxford.

Mikus, B. (2001). *Risikomanagement*, Chapter Risiken und Risikomanagement, pp. 3–28.

Miles, R., & Snow, C. (1978). *Organizational strategy, structure, and process*. New York: McGraw-Hill.

Milgram, S. (1967). The small world problem. *Psychology Today, 2*, 60–67.

Milgrom, P., & Roberts, J. (1990). Rationalizability, learning, and equilibrium in games with strategic complementarities. *Econometrica, 58*, 1255–1277.

Mizruchi, M. (1982). *The American Corporate Network*. Beverley Hills: Sage.

Möller, H. (1985). Die Informationseffizienz des deutschen Kapitalmarktes. *Zeitschrift für betriebswirtschaftliche Forschung, 37*, 500–518.

Monk, E., & Wagner, B. (2006). *Concepts in Enterprise Resource Planning*. Boston: Thomson Course Technology.

Montoya, J., & Sole, R. (2002). Small world patterns in food webs. *Journal of Theoretical Biology, 214*, 405–412.

Moreno, J. (1934). *Who Shall Survive?* Beacon: Beacon House.

Mossin, J. (1966). Equilibrium in a capital asset market. *Econometrica, 34*, 768–783.

Moxter, A. (1991). *Grundsätze ordnungsmäßiger Unternehmensbewertung*. Wiesbaden: Gabler.

Mueller, J., Schoenfisch, B., & Kirkilionis, M. (2000). Ring vaccination. *Journal of Mathematical Biology, 41*, 143–171.

Mueller, S. (2003). *Die Bewertung junger Unternehmen und Behavioral Finance - Eine theoretische und experimentelle Untersuchung*. Universität Münster.

Mulhern, F. (1999). Customer profitability analysis: measurement, concentration, and research directions. *Journal of Interactive Marketing, 13*, 25–40.

Müller, M. (1990). *Software-Unternehmen am deutschen Software-Markt: Leistungswirtschaftliche Besonderheiten, Wettbewerbsbedingungen und Gestaltungsmaßnahmen*. Düsseldorf: VDI.

Müller, S. (2003). *Die Bewertung junger Unternehmen und Behavioral Finance: Eine theoretische und experimentelle Untersuchung*. Lohmar: Josef Eul.

Muthsam, H. (2006). *Lineare Algebra und ihre Anwendungen*. München: Elsevier.

Myers, S. (1977). Determinants of corporate borrowing. *Journal of Financial Economics, 5*, 147–175.

Naisbitt, J., Naisbitt, N., & Philips, D. (2001). *High tech high touch: Technology and Our Accelerated Search for Meaning*. London: Nicholas Brealey.

Neftci, S. (1996). *An Introduction to the Mathematics of Financial Derivatives*. New York: Academic.
Neftci, S. (2000). *An Introduction to the Mathematics of Financial Derivatives* (2nd ed.). New York: Academic Press.
Neugebauer, U., Marock, J., & Bujara, G. (1983). *Der Markt fuer Software Systeme und DV-bezogene Dienstleistungen in der Bundesrepublik Deutschland*. St. Augustin: GMD Forschungszentrum Informationstechnik, Sankt Augustin.
Newman, M. (2000). Models of the small world. *Journal of statistical physics, 101*, 819–841.
Newman, M. (2001a). Clustering and preferential attachment in growing networks. *Physical Review E, 64*, 025102.
Newman, M. (2001b). The structure of scientific collaboration networks. *Proceedings of the National Academy of Sciences of the United States of America, 98*, 404–409.
Newman, M. (2002). Assortative mixing in networks. *Physical Review Letters, 89*, 208701.
Newman, M. (2003a). Mixing patterns in networks. *Physical Review E, 67*(2), 026126.
Newman, M. (2003b). The structure and function of complex networks. *SIAM Review, 45*, 167–256.
Newman, M., Barabási, A., & Watts, D. (2006). *The structure and dynamics of networks*. NJ: Princeton University Press.
Newman, M., Forrest, S., & Balthrop, J. (2002). Email networks and the spread of computer viruses. *Physical Review E, 66*, 035101.
Newman, M., & Watts, D. (1999). Scaling and percolation in the small-world network model. *Physical Review E, 60*, 7332–7342.
Niederdrenk, R., & Maack, B. (2008). Trumpf im Bieterwettbewerb: Die richtige Commercial Due Diligence. *M&A Review, 5*, 229–233.
Niraj, R., Gupta, M., & Narasimham, C. (2001). Customer profitability in a supply chain. *Journal of Marketing, 65*, 25–40.
Ohashi, H. (2003, December). The Role of Network Effects in the US VCR Market, 1978–1986. *Journal of Economics & Management Strategy, 12*(4), 447–494.
Parkinson, M. (1977). Option pricing: The american put. *The Journal of Business, 50*(1), 21–36.
Pastor-Satorras, R., Vazquez, A., & Vespignani, A. (2001). Dynamical and correlation properties of the internet. *Physical Review Letters, 87*, 258701.
Pastor-Satorras, R., & Vespignani, A. (2001a). Epidemic dynamics and endemic states in complex networks. *Physical Review E, 63*, 066117.
Pastor-Satorras, R., & Vespignani, A. (2001b). Epidemic spreading in scale-free networks. *Physical Review Letters, 86*, 3200–3203.
Pastor-Satorras, R., & Vespignani, A. (2002). Immunization of complex networks. *Physical Review E, 65*, 036104.
Pastor-Satorras, R., & Vespignani, A. (2003). *Handbook of Graphs and Networks*, Chapter Epidemics and immunization in scale-free networks. Berlin: Wiley-VCH.
Payne, A., Holt, S., & Frow, P. (2001). Relationship value management. *Journal of Marketing Management, 17*, 785–817.
Pechtl, H. (2005). *Preispolitik*. Stuttgart: Lucius & Lucius.
Peemöller, V., Beckmann, C., & Kronmüller, A. (2002). Empirische Erhebung zum aktuellen Stand der praktischen Anwendung des Realoptionsansatzes. *Finanz Betrieb, 4*, 197–209.
Peemöller, V., & Kunowski, S. (2002). *Praxishandbuch der Unternehmensbewertung*, Chapter Ertragswertverfahren nach dem IDW.
Peemoeller, V., Boemelburg, P., & Denkmann, A. (1994). Unternehmensbewertung in Deutschland: Eine empirische Erhebung. *Die Wirtschaftsprüfung, 22*, 741–749.
Pennock, D., Flake, G., Lawrence, S., Glover, E., & Giles, C. (2002). Winners don't take all: Characterizing the competition for links on the web. *Proceedings of the National Academy of Sciences of the United States of America, 99*, 5207Ű5211.
Pólya, G. (1937). Kombinatorische Anzahlbestimmungen für Gruppen, Graphen und chemische Verbindungen. *Acta Math., 68*, 145–254.

Porter, M. (2001). Strategy and the Internet. *Harvard Business Review, 79*(winter special Issue), 63–78.
Porter, M. (2008). The five competitive forces that shape strategy. *Harvard Business Review, 86*, 78–93.
Prahalad, C., & Hamel, G. (1990). The core competencies of the corporation. *Harvard Business Review, 68*(3), 79–91.
Price, D. (1965). Networks of scientific papers. *Science, 149*, 510–515.
Price, D. (1976). A general theory of bibliometric and other cumulative advantage processes. *Journal of the American Society for Information Science, 27*, 292–306.
Rendleman, R. J., & Bartter, B. (1979). Two-state option pricing. *Journal of Finance, 34*, 1093–1110.
Ramge, T. (2008). Von Freund zu Freund - Was früher Unternehmen vorbehalten war, können Kunden heute selbst: Marken machen. *Brand eins, 2*, 68–73.
Rapoport, A. (1957). Contribution to the theory of random and biased nets. *Bulletin of Mathematical Biophysics, 19*, 257–277.
Rayport, J. (1996). The virus of marketing. *Fast Company, 6*, 68–71.
Röder, K., & Müller, S. (2001). Mehrperiodige Anwendung des CAPM im Rahmen von DCF-Verfahren. *Finanz Betrieb, 3*, 225–233.
Regibeau, P., & Rockett, K. (1996). The timing of product introduction and the credibility of compatibility decisions. *International Journal of Industrial Organization, 14*, 801–823.
Reichheld, F. (1999). *The Loyalty Effect*. Boston: Bain, The Free Press.
Reichheld, F., & Sasser, E. (1990). Zero defection: quality comes to service. *Harvard Business Review, 68*(5), 2–8.
Reinartz, W., & Kumar, V. (2000). On the profitability of long-life customers in a noncontractual setting: an empirical investigation and implications for marketing. *Journal of Marketing, 64*, 17–35.
Richins, M. (1983). Negative word-of-mouth by dissatisfied consumers: A pilot study. *Journal of Marketing, 47*, 68–78.
Rogers, E. (1983). *Diffusion of innovations*. New York: The Free Press.
Rogers, E. (1995). *Diffusion of innovations*. New York: The Free Press.
Rogers, E. (1996). *Diffusion of Innovations* (4th ed.). New York: The Free Press.
Rogers, E., & Beal, G. (1958). The importance of personal influence in adoption of technological changes. *Social Forces, 36*(4), 329–334.
Rohjahn, J., & Berner, C. (2002). Wertorientiertes Management von Wachstumsunternehmen unter Berücksichtigung des Realoptionsansatzes. *Finanz Betrieb, 4*, 1–6.
Rohlfs, J. (1974). A theory of interdependent demand for a communication service. *Bell Journal of Economics and Management Science, 5*, 16–37.
Roll, R. (1977). Critique of the asset pricing theory's tests. *Journal of Financial Economics, 5*, 129–176.
Rosenblueth, A., Wiener, N., & Bigelow, J. (1943). Behavior, purpose and teleology. *Philosophy of Science, 10*, 18–24.
Ross, S. (1977). The arbitrage theory of capital asset pricing. *Journal of Economic Theory, 13*, 341–360.
Ross, S., Westerfield, R., & Jaffe, J. (1996). *Corporate Finance*. Boston MA: McGraw-Hill.
Rust, R., Zahorik, A., & Keiningham, T. (1995). Return on quality (roq): Making service quality financially accountable. *Journal of Marketing, 59 (April)*, 58–70.
Salop, S., & Scheffman, D. (1983). Raising rivals' cost. *American Economic Review Papers and Proceedings, 73*, 267–271.
Sander, L., Warren, C., Sokolov, I., Simon, C., & Koopman, J. (2002). Percolation on disordered networks as a model for epidemics. *Mathematical Biosciences, 180*, 293–305.
Sattenspiel, L., & Simon, C. (1988). The spread and persistence of infectious diseases in structured populations. *Mathematical Biosciences 90*, 367–383.

Schäfer, H., & Schässburger, B. (2001). Bewertungsmängel von CAPM und DCF bei innovativen wachstumsstarken Unternehmen und optionspreistheoretische Alternativen. *Zeitschrift für Betriebswirtschaft, 71*, 85–107.

Scherer, F., Harhoff, D., & Kukies, J. (2000). Uncertainty and the size distribution of rewards from innovation. *Journal of Evolutionary Economics, 10*, 175–200.

Schilling, M. (1999). Winning the standards race: Building installed base and the availability of complementary goods. *European Management Journal, 17*, 265–274.

Schluechtermann, J. (1996). *Multiagent Engineering*, Chapter Planung in zeitlich offenen Entscheidungsfeldern. Heidelberg: Springer.

Schmalen, H., Binninger, F., & Pechtl, H. (1993). Diffusionsmodelle als Entscheidungshilfe zur Planung absatzpolitischer Maßnahmen bei Neuprodukteinführungen. *Die Betriebswirtschaft, 53*, 513–527.

Schmittlein, D., Morrison, D., & Colombo, R. (1987). Counting your customers: Who are they and what will they do next? *Management Science, 33*, 1–24.

Schmittlein, D., & Peterson, R. (1994). Customer base analysis. *Marketing Science, 13*, 41–67.

Schoder, D. (1995). *Erfolg und Misserfolg telematischer Innovationen*. Wiesbaden.

Schulmerich, M. (2005). *Real Options Valuation - The Importance of Interest Rate Modelling in Theory and Practice*. Berlin: Springer.

Schwartz, E., & Moon, M. (2000). Rational pricing of internet companies. *Financial Analysts Journal, 56*, 62–75.

Schwartz, N., Cohen, R., ben Avraham, D., Barabási, A., & Havlin, S. (2002). Percolation in directed scale-free networks. *Physical Review E, 66*, 015104.

Scott, J. (2000). *Social Network Analysis: A Handbook* (2nd ed.). London: Sage Publications.

Sebastian, K., Niederdrenk, R., & Tesch, A. (2005). *Due Diligence bei Unternehmensakquisitionen* (4 ed.)., Chapter Market Due Diligence. Stuttgart: Schaeffer-Poeschel.

Segars, A., & Grover, V. (1993). Re-examining perceived ease of use and usefulness: A confirmatory factor analysis. *MIS Quarterly, 17*, 517–525.

Shankar, R. (1994). *Principles of Quantum Mechanics*. Berlin: Springer.

Shapiro, C., & Varian, H. (1998). *Information Rules: A Strategic Guide to the Network Economy*. Harvard: Harvard Business School.

Shapiro, C., & Varian, H. (1999). The art of standard wars. *California Management Review, 41*(2), 8–32.

Sharpe, W. (1964). Capital asset prices: a theory of market equilibrium under conditions of risk. *Journal of Finance, 19*, 425–442.

Sheth, J., & Parvatiyar, A. (1995). Relationship marketing in consumer markets: Antecedents and consequences. *Journal of the Academy of Marketing Science, 23*, 255–271.

Shiller, R. (2000). *Irrational Exuberance*. Princeton: Princeton University Press.

Shleifer, A. (2000). *Inefficient Markets - An Introduction to Behavioral Finance*. New York: Oxford University Press.

Shy, O. (2000a). *Industrial Organization: theory and application*. Cambridge: MIT.

Shy, O. (2000b). *Network Effects in Economics*.

Shy, O. (2001). *The Economics of Network Industries*. Cambridge: Oxford University Press.

Sick, G. (1989). *Capital budgeting with real options*. New York: Salomon Brothers Center for the Study of Financieal Institutions.

Simon, H. (1955). On a class of skew distribution functions. *Biometrika, 42*, 425–440.

Simon, H. (1962). *The architecture of complexity*. Cambridge: MIT. Fundamental Article.

Skiera, B. (1999). *Mengenbezogene Preisdifferenzierung bei Dienstleistungen*. Wiesbaden: Gabler DUV.

Solomon, M., & Rabolt, N. (2004). *Consumer behavior: in fashion*. New Jersey: Prentice-Hall.

Solomonoff, R., & Rapoport, A. (1951). Connectivity of random nets. *Bulletin of Mathematical Biophysics, 13*, 107–117.

Späth, G. (1994). *Preisstrategien für innovative Telekommunikationsleistungen*. Wiesbaden: Gabler.

Srivastava, R., Shervani, T., & Fahey, L. (1998). Marked-based assets and shareholder value: A framework analysis. *Journal of Marketing, 62*, 2–18.

Srivastava, R., Shervani, T., & Fahey, L. (1999). Marketing, business processes, and shareholder value. *Journal of Marketing, 63*(4), 168–179.

Staat, M., Bauer, H., & Hammerschmidt, M. (2002). Structuring product-markets: an approach based on customer value. *Marketing Theory and Applications, 13*, 205–212.

Stemmann, M., & Treptow, G. (2001). Empirische Analyse von Unternehmen der New Economy vor dem Hintergrund des Technology-Builder Modells von Price Waterhouse. *Finanz Betrieb, 3*, 404–410.

Sterman, J. (2000). *Business Dynamics: Systems Thinking and Modeling for a Complex World*. Boston: Irwin/McGraw-Hill.

Strahringer, S., & Westner, M. (2008). Evaluation criteria for selecting offshoring candidates: An analysis of practices in german businesses. *Journal of Information Technology Management, 19*, 19–34.

Strang, G. (2003). *Lineare Algebra*. Berlin: Springer.

Strogatz, S. (1994). *Nonlinear Dynamics and Chaos*. Cambridge: Perseus.

Strogatz, S. (2001). Exploring complex networks. *Nature, 410*, 268–276. Nonlinear Dynamics; Regular Networks of coupled dynamical systems: identical oscilators, non-identical oscilators; Complex Networks architecture: Random, Small-World, Scale-Free networks, Random Graphs; Outlook.

Strogatz, S. (2004). *Sync - Rhythms of Nature, Rhythms of Ourselves*. Baltimore, MD: Penguin Books.

Stulz, R. (1982). Options on the minimum or the maximum of two risky assets. *Journal of Financial Economics, 10*, 161–185.

Sturm, S. (2003). *Bewertung von Unternehmen: Strategie - Markt - Risiko*, Chapter Bewertung von geschichtslosen Unternehmen, pp. 207–234.

Subramanian, G. (1994). A replication of perceived usefulness and perceived ease of use measurement. *Decision Sciences, 25*, 863–873.

Sullivan, K., Chalasani, P., Jha, S., & Sazawal, V. (1999). Software Design as an Investment Activity: A Real Options Perspective. In L. Trigeorgis (Ed.), *Real Options and Business Srategy: Applications to Decision Making* (pp. 215–261). London: Risk Books.

Swann, P., & Shurmer, M. (1994). The emergence of standards in pc software: who would benefit from institutional intervention? *Information Economics and Policy, 6*, 295–318.

Taudes, A. (1998). Software growth options. *Journal of Management Information Systems, 15*, 165–185.

Taudes, A., Feuerstein, M., & Mild, A. (2000). Options analysis of software platform decisions: A case study. *MIS Quarterly, 24*, 227–243.

Tirole, J. (1999). *Industrieökonomik*. München: Oldenbourg.

Topkis, D. (1979). Equilibrium points in nonzero-sum n-person submodular games. *SIAM Journal of Control and Optimization, 17*, 773–787.

Tornatzky, L., & Klein, R. (1982). Innovation characteristics and innovation adoption-implementation: A meta-analysis of findings. *IEEE Transactions on Engineering Management, 29*, 28–45.

Travers, J., & Milgram, S. (1969). An experimental study of the small world problem. *Sociometry, 32*, 425–443.

Trigeorgis, L. (1991). A log-transformed binomial numerical analysis method for valuing complex multi-option investments. *Journal of Financial and Quantitative Analysis, 26*, 309–326.

Trigeorgis, L. (1996). *Real options: Managerial flexibility and strategy in resource allocation*. Boston: MIT.

Vahlen (2006). *Vahlens großes Wirtschaftslexikon*, Chapter Preisstrategien. München: Vahlen.

Valente, T. (1995). *Network Models of the Diffusion of Innovations*. Cresskill: Hampton.

Valente, T., & Davis, R. (1999). Accelerating the diffusion of innovations using opinion leaders. *Annals of the American Academy, 566*, 55–67.

Varian, H. (1993). *Economic and Financial Modeling with Mathematica*. New York: Springer.

Vazquez, F., Krapivsky, P., & Redner, S. (2003). Constrained opinion dynamics: Freezing and slow evolution. *Journal of Physics A, 36*, L61–L68.

VDI (1983). Technical report, Verband deutscher Industrie - Technologie Zentrum.

Veblen, T. (1899). *The theory of the leisure class: An economic study of institutions*. Boston: Houghton Mifflin.

Venkatesh, V., & Davis, F. (2000). A theoretical extension of the technology acceptance model: Four longitudinal field studies. *Management Science, 46*(2), 186–204.

Verdult, V., & Verhaegen, M. (2000). Bilinear state space systems for nonlinear dynamical modelling. *Theory Biosciences, 119*, 1–9.

Vives, X. (1999). *Oligopoly Pricing. Old Ideas and New Tools*. Cambridge, MA: MIT.

Vollrath, R. (2001). *Realoptionen in der Unternehmenspraxis. Werte schaffen durch Fleibilität*, Chapter Die Berücksichtigung von Handlungsflexibilität bei Investitionsentscheidungen - Eine empirische Untersuchung. Berlin: Springer.

von Westarp, F. (2003). *Modeling Software Markets - Empirical Analysis, Network Simulations, and Marketing Implications*. Heidelberg: Springer.

Wamser, C. (2000). *Electronic Commerce - Grundlagen und Perspektiven*, Chapter Electronic Commerce - theoretische Grundlagen und praktische Relevanz, pp. 3–27. München.

Wang, P., & Spiegel, T. (1994). Database marketing and its measurement of success: designing a managerial instrument to calculate the value of a repeat customer base. *Journal of Direct Marketing, 8*, 73–81.

Warren, C., & Sokolov, L. S. I. (2002). Geography in a scale-free network model. *Physical Review E, 66*, 056105.

Wasserman, S., & Faust, K. (1994). *Social network theory*. Cambridge: Cambridge University Press.

Watts, D. (1999). *Small worlds: the dynamics of networks between order and randomness*. Princeton: Princeton University Press.

Watts, D., & Preretti, J. (2007). Viral marketing for the real world. *Harvard Business Review, 5*, 22–23.

Watts, D., & Strogatz, S. (1998). Collective dynamics of 'small-world' networks. *Nature, 393*, 440–442.

Weiber, R. (1993). Chaos: Das Ende der klassischen Diffusionsmodellierung? *Marketing Zeitschrift für Forschung und Praxis, 19*, 35–46.

Weitzel, T. (2004). *Economics of Standards in Information Networks*. Heidelberg: Springer.

Weitzel, T., Beimborn, D., & König, W. (2002). An individual view on cooperation networks. *Proceedings of the 36th Hawaii International Conference on System Sciences*.

Weitzel, T., Wendt, O., & Westarp, F. (2000). Reconsidering network effect theory. *European Conference on Information Systems 2000*.

Welt (2008). Toshiba beerdigt die hoch auflösende HD-DVD. *Die Welt February 19th*.

Wied-Nebbeling, S. (2003). *Preistheorie und Industrieökonomik*. Berlin: Springer. Price Theory, Credence Goods, Network Goods, Industry Economics, gut zu lesen.

Wiese, H. (1990). *Netzeffekte und Kompatibilität - Ein theoretischer und simulationsgeleiteter Beitrag zur Absatzpolitik für Netzeffekt-Güter*. Stuttgart: C.E. Poeschel Verlag.

Wiese, H. (1991). Marktschaffung: Das startproblem bei netzeffekt-gütern. *Marketing ZFP, 1*, 43–51.

Williams, R., & Martinez, N. (2000). Simple rules yield complex food webs. *Nature, 404*, 180–183.

Willner, R. (1995). Valuing start-up venture growth options. In L. Trigeorgis (Ed.), *Real Options in Capital Investment - Models, Strategie and Applications* (pp. 221–239). Westport: Praeger.

Wilmott, P. (2000). *Wilmott on Quantitative Finance*. New York: John Wiley & Sons.

Wilmott, P., Howison, S., & Sewyenne, J. (1995). *The Mathematics of Financial Derivatives*. Cambridge UK: Cambridge University Press. Basic Option Theory, Numerical Methods, Further Option Theory.

Wirtz, J., & Kam, W. (2001). An empirical study on internet-based business-to-business ecommerce in singapore. *National University of Singapore, 1*, 1–30.

Xing (2007a). *Annual Report 2007*. Xing AG.

Xing (2007b). Kommunikation und Networking im Internet: 3. Internationale XING-Studie. *Xing AG.*

Yang, Y. (1996). Network effects, pricing strategies, and optimal upgrade time in software provision. *EconWPA, Industrial Organization*, 9602001.

Yule, G. (1925). A mathematical theory of evolution, based on the conclusions of dr. j. c. willis, f.r.s. *Philosophical Transactions of the Royal Society of London, Series B, 213*, 21–87.

Zanini, M. (2005). Strategy in an era of global giants. *The McKinsey Quarterly, 4*, 47–59.

Zanini, M. (2008). Using power curves to assess industry dynamics. *The McKinsey Quarterly, November*, 1–5.

Zerr, K. (2004). *Applied Marketing: Anwendungsorientierte Marketingwissenschaft der deutschen Fachhochschulen*, Chapter Guerilla Marketing in der Kommunikation: Kennzeichen, Mechanismen und Gefahren, pp. 583–590. Springer, Berlin.

Index of Keywords

FSC
www.fsc.org
MIX
Papier | Fördert
gute Waldnutzung
FSC® C083411

Kolibri 360 GmbH
Ferdinand-Jühlke-Straße 7
99095 Erfurt, Deutschland
produktsicherheit@kolibri360.de